T0205547

Studies in Computational Intelligence

Volume 788

Series editor

Janusz Kacprzyk, Polish Academy of Sciences, Warsaw, Poland
e-mail: kacprzyk@ibspan.waw.pl

The series "Studies in Computational Intelligence" (SCI) publishes new developments and advances in the various areas of computational intelligence—quickly and with a high quality. The intent is to cover the theory, applications, and design methods of computational intelligence, as embedded in the fields of engineering, computer science, physics and life sciences, as well as the methodologies behind them. The series contains monographs, lecture notes and edited volumes in computational intelligence spanning the areas of neural networks, connectionist systems, genetic algorithms, evolutionary computation, artificial intelligence, cellular automata, self-organizing systems, soft computing, fuzzy systems, and hybrid intelligent systems. Of particular value to both the contributors and the readership are the short publication timeframe and the world-wide distribution, which enable both wide and rapid dissemination of research output.

More information about this series at http://www.springer.com/series/7092

Roger Lee

Editor

Applied Computing and Information Technology

 Springer

Editor
Roger Lee
Software Engineering and Information
    Technology Institute
Central Michigan University
Mount Pleasant, MI, USA

ISSN 1860-949X   ISSN 1860-9503   (electronic)
Studies in Computational Intelligence
ISBN 978-3-030-07489-0   ISBN 978-3-319-98370-7   (eBook)
https://doi.org/10.1007/978-3-319-98370-7

This Springer imprint is published by the registered company Springer Nature Switzerland AG
The registered company address is: Gewerbestrasse 11, 6330 Cham, Switzerland

Foreword

The purpose of the 6th International Conference on Applied Computing and Information Technology (ACIT 2018) held on June 13–15, 2018, in Kunming, China, was to together researchers, scientists, engineers, industry practitioners, and students to discuss, encourage, and exchange new ideas, research results, and experiences on all aspects of applied computers and information technology, and to discuss the practical challenges encountered along the way and the solutions adopted to solve them. The conference organizers have selected the best 13 papers from those papers accepted for presentation at the conference in order to publish them in this volume. The papers were chosen based on review scores submitted by members of the program committee and underwent further rigorous rounds of review.

In chapter 1, Soo-Yeon Yoon, Yang-Ha Chun, Jong-Bae Kim conducted a server to verify the relationship between quality factors, expected value, and educational satisfaction of NCS education system for university students. The study suggested implications for the satisfaction of the values and educational satisfaction of the students in terms of educational service quality provided by the NCS education system.

In chapter 2, Veiga Yamodo Marland and Haeng-Kon Kim present a way to simplify the creation of applications for mobile platforms by developing a high-level and platform-independent model of an application, and automatically transforming this high-level model to platform-specific code.

In chapter 3, Mechelle Grace Zaragoza, Haeng-Kon Kim, and Ha Jin Hwang present a study that shows how mobile learning can transform the delivery of education and training. This study is timely because there is significant growth in the use of mobile technology by people around the world, especially in developing countries.

In chapter 4, Symphorien Karl Yoki Donzia, Hyun Yeo, and Haeng-Kon Kim present a study that includes the expansion of rice management scales, including human resources, water and agricultural land; response strategies to irrigation during the drought; and the application of several agricultural water resources. Finally, it presents a architecture for IoT agriculture services based on cloud sensor infrastructure. Their results show that rice should be planted in lowland rice production systems in Central Africa.

In chapter 5, Kiljae Ahn, Dae-Sik Ko, and Sang-Hoon Gim introduce the mixed reality development in architectural design and construction business and suggest design collaboration service platform composed of mixed reality service server, which is a Web service system for authoring and managing mixed reality contents, and terminal application that implements it in mixed reality environment.

In chapter 6, Donghyuk Jo presents a study of the information communication technology intended to suggest Defense Information System (C4I) success model to improve the operational performance of the C4I system and to empirically verify the model.

In chapter 7, Lamya Alqaydi, Chan Yeob Yeun, and Ernesto Damiani provide an overview of the landscape of vulnerabilities relating to SSL/TLS protocol versions with estimated risk levels. Selecting the best configuration for a given use case is a time-consuming task, and testing a given configuration of a server for all known vulnerabilities is also difficult.

In chapter 8, Jeong Hoon Shin and Hyo Won Jeon analyze the patterns of variation of the cerebral activation status, depending on blood types, in response to external stimulation. The results of this study would provide the basis for classification of subjects, which is necessary for the selection of effective auditory stimulant sound sources in the process of neurofeedback therapy/training, inducing the users to modify their own brain waves for healing, and would also present the measures for the selection of efficient visual and auditory stimuli.

In chapter 9, Kyung-Jin Jung, Jung-Boem Park, Nhu Quynh Phan, Chen Bo, and Gwang-yong Gim present research that seeks to figure how characteristics among countries respond to the intention of usage for cryptocurrencies by comparing Korea, China, and Vietnam.

In chapter 10, Ki-Seob Hong, Hyo-Bin Kim, Dong-Hyun Kim, and Jung-Taek Seo present a study that suggests a technology that can detect attacks exhibiting a replay pattern against ICS, using white list-based detection and machine learning to educate control traffic and apply the results to actual detection.

In chapter 11, Benedicto B. Balilo Jr., Bobby D. Gerardo, and Yungcheol Byun present the new algorithm based on mixed technique by substituting reverse scheme, swapping, and bit-manipulation. This encryption technique is a concept that combines selected cryptographic technique that can reduce and save time processing encryption operation, offers simplicity, and extended mixed encryption key generation.

In chapter 12, Eric Goold, Sean O'Neill, and Gongzhu Hu proposed a new clustering algorithm, called reciprocating link hierarchical clustering, which considers the neighborhood of the points in the data set in term of their reciprocating affinity, while accommodating the agglomerative hierarchical clustering paradigm.

In chapter 13, Mechelle Grace Zaragoza and Haeng-Kon Kim develop an Infant Care System Application. They discuss two main topics: the first topic is related to SIDS vulnerability, to better understand the role of learning during sleep in promoting infant survival; the second one related the importance of making sure that infants are given the comfort they need in order to continuously sleep without disturbing the guardians.

It is our sincere hope that this volume provides stimulation and inspiration, and that it will be used as a foundation for works to come.

July 2018   Takaaki Goto

Contents

Factors Affecting Satisfaction of NCS Based Educational System 1
Soo-Yeon Yoon, Yang-Ha Chun, and Jong-Bae Kim

**Model-Driven Development of Mobile Applications Allowing
Role-Driven Variants** 14
Veiga Yamodo Marland and Haeng-Kon Kim

E-Learning Adaptation and Mobile Learning for Education 27
Mechelle Grace Zaragoza, Haeng-Kon Kim, and Ha Jin Hwang

**Design and Evaluation of Soil pH IoT Sensor Attribute for Rice
Agriculture in Central Africa** 37
Symphorien Karl Yoki Donzia, Hyun Yeo, and Haeng-Kon Kim

**A Study on the Architecture of Mixed Reality Application
for Architectural Design Collaboration** 48
Kiljae Ahn, Dae-Sik Ko, and Sang-Hoon Gim

Exploring the Improvement of the Defense Information System 62
Donghyuk Jo

**A Modern Solution for Identifying, Monitoring, and Selecting
Configurations for SSL/TLS Deployment** 78
Lamya Alqaydi, Chan Yeob Yeun, and Ernesto Damiani

**Analyses of Characteristics of Changes in Cerebral Activation Status,
Depending on Blood Types, in Response to Auditory Stimulation** 89
Jeong Hoon Shin and Hyo Won Jeon

**An International Comparative Study on the Intension
to Using Crypto-Currency** 104
Kyung-Jin Jung, Jung-Boem Park, Nhu Quynh Phan, Chen Bo,
and Gwang-yong Gim

Detection of Replay Attack Traffic in ICS Network 124
Ki-Seob Hong, Hyo-Bin Kim, Dong-Hyun Kim, and Jung-Taek Seo

CipherBit192: Encryption Technique for Securing Data 137
Benedicto B. Balilo Jr., Bobby D. Gerardo, and Yungcheol Byun

Reciprocating Link Hierarchical Clustering 149
Eric Goold, Sean O'Neill, and Gongzhu Hu

Development of Infant Care System Application 166
Mechelle Grace Zaragoza and Haeng-Kon Kim

Author Index 175

List of Contributors

Kiljae Ahn Mokwon University, Daejeon, Republic of Korea

Lamya Alqaydi ECE Department, Khalifa University of Science and Technology, Abu Dhabi, UAE

Benedicto B. Balilo Jr. CSIT Department, Bicol University Legazpi City, Albay, Philippines

Chen Bo Business Administration, Soongsil University, Seoul, Korea

Yungcheol Byun Department of Computer Engineering, Jeju National University, Jeju, Korea

Yang-Ha Chun Department of Computer Science, Yongin University, Gyeonggi-do, Korea

Ernesto Damiani ECE Department, Khalifa University of Science and Technology, Abu Dhabi, UAE

Symphorien Karl Yoki Donzia Daegu Catholic University, Gyeongsan, South Korea

Bobby D. Gerardo Institute of ICT, West Visayas State University, Lapaz Iloilo City, Philippines

Gwang-yong Gim Business Administration, Soongsil University, Seoul, Korea

Sang-Hoon Gim Dongwoo E&C co., Seoul, Republic of Korea

Eric Goold Department of Computer Science, Central Michigan University, Mount Pleasant, MI, USA

Ki-Seob Hong Department of Information Security Engineering, Soonchunhyang University, Asan, South Korea

Gongzhu Hu Department of Computer Science, Central Michigan University, Mount Pleasant, MI, USA

Ha Jin Hwang Sunway University Business School, Sunway University, Subang Jaya, Malaysia

Hyo Won Jeon Department of IT Engineering, Daegu Catholic University, Gyeongsan-si, Gyeongbuk, Republic of Korea

Donghyuk Jo Soongsil University, Seoul, South Korea

Kyung-Jin Jung Business Administration, Soongsil University, Seoul, Korea

Dong-Hyun Kim Department of Information Security Engineering, Soonchunhyang University, Asan, South Korea

Haeng-Kon Kim School of Information Technology, Daegu Catholic University, Gyeongsan, Gyeongsangbuk-do, South Korea

Hyo-Bin Kim Department of Information Security Engineering, Soonchunhyang University, Asan, South Korea

Jong-Bae Kim Graduate School of Software, Soongsil University, Seoul, Korea

Dae-Sik Ko Mokwon University, Daejeon, Republic of Korea

Veiga Yamodo Marland Computer Information Engineering, Daegu Catholic University, Gyeongsan-si, Gyeongsangbuk-do, South Korea

Sean O'Neill Department of Computer Science, Central Michigan University, Mount Pleasant, MI, USA

Jung-Boem Park Business Administration, Soongsil University, Seoul, Korea

Nhu Quynh Phan Business Administration, Soongsil University, Seoul, Korea

Jung-Taek Seo Department of Information Security Engineering, Soonchunhyang University, Asan, South Korea

Jeong Hoon Shin Department of IT Engineering, Daegu Catholic University, Gyeongsan-si, Gyeongbuk, Republic of Korea

Hyun Yeo Sun Cheon National University, Suncheon, Korea

Chan Yeob Yeun ECE Department, Khalifa University of Science and Technology, Abu Dhabi, UAE

Soo-Yeon Yoon Department of Faculty of Arts, Soongsil University, Seoul, Korea

Mechelle Grace Zaragoza School of Information Technology, Daegu Catholic University, Gyeongsan, South Korea

Factors Affecting Satisfaction of NCS Based Educational System

Soo-Yeon Yoon[1], Yang-Ha Chun[2], and Jong-Bae Kim[3(✉)]

[1] Department of Faculty of Arts, Soongsil University, Seoul, Korea
ll04py@naver.com
[2] Department of Computer Science, Yongin University, Gyeonggi-do, Korea
yanghal000@naver.com
[3] Graduate School of Software, Soongsil University, Seoul, Korea
kjbl23@ssu.ac.kr

Abstract. The purpose of this study is to verify the relationship between quality factors, expected value and educational satisfaction of NCS education system for university students A total of 289 college students from Seoul and Gyeonggi provinces were selected for the study and a questionnaire survey was conducted from March to April 2017. The collected questionnaires were analyzed using SPSS 22.0 and AMOS 22.0 statistical programs and the following study results were derived. First, responsibility, contentability, assurance and usefulness have significant influence on perceived value, and reactivity and tangibility have no effect on quality factor of NCS education system. Second, responsibility, contentability, assurance and usefulness have a significant effect on educational satisfaction, and reactivity and tangibility have no effect on educational satisfaction. Finally, the perceived value of the NCS education system has a significant effect on educational satisfaction. Therefore, this study suggested implications for the satisfaction of the values and educational satisfaction of the students in terms of educational service quality provided by the NCS education system.

Keywords: NCS · Education system · Education service
Education satisfaction

1 Introduction

Recently, 'National Competency Standards' (NCS) have been promoted as one of the national tasks for the implementation of capacity-based society. National Competency Standards (NCS) began to be developed in 2002 and have been systematized and developed for many years since then. It has been introduced and operated by vocational education institutions such as Polytechnics, Technical Education Universities, and Specialized and Specialized High Schools for several years.

From 2015, general professional vocational education institutions will be used to certify NCS professional education institutions. In addition, the government is also encouraging the introduction of NCS-based curriculum to general higher education institutions. In the industrial sector, the curriculum of higher vocational education institutions is required to be reorganized into a job-oriented curriculum.

R. Lee (Ed.): ACIT 2018, SCI 788, pp. 1–13, 2019.
https://doi.org/10.1007/978-3-319-98370-7_1

Therefore, the government is demanding the operation of NCS - based curriculum for higher vocational education institutions. In addition, the college is revising its major curriculum based on NCS in order to solve the inconsistency between the education site and the industrial site in accordance with the government's demand. In order to improve the NCS - based education and administrative service quality of colleges, we applied the SERVQUAL theory to the NCS education system.

We analyzed the quality of NCS-based educational services and the environmental service quality that constitute them in the college, and examined whether they affect perceived value and educational satisfaction. The purpose of this study is to provide academic and practical implications for environmental and social job utilization of NCS - based curriculum.

2   Related Research

2.1   National Competency Standards

National Competency Standards (NCS) are guidelines or regulations that are agreed to by consensus as mutual promises to promote convenience and efficiency among stakeholders [2]. In this definition, NCS is intended to develop development teams such as the Industrial Human Resources Development Council (SC), which are composed of training/qualification and field/education experts, to be used for training courses and qualification standards.

According to Article 2 of the <Qualification Framework Act>, NCS defines "the ability (knowledge, skill, attitude) necessary to successfully perform the job in the industrial field to be standardized at national level".

NCS aims to improve the quality of education and training so that it reaches the level of job ability suggested by NCS, so as to train and supply manpower to meet the demand of industrial field.

2.2   Formulas

The NCS-based curriculum was developed by developing a "study guide" for trainees and a "training operation plan" for teachers and instructors with specific learning plans and learning methods and evaluations of education and training applying NCS-based training standards Process.

The training operational plan presents a level of training completion level that meets the level of curriculum and curriculum proposed in the NCS-based training standard.

The NCS - based training curriculum and education subject name reflected in the training subject summary table presents the classification number for each competency unit element.

The education and training plan for each subject includes all the performance criteria presented in the curriculum (ability unit) of the curriculum corresponding to the NCS-based training standard, and presents appropriate performance evaluation and evaluation methods.

The outline of the training course in the study guide, the study guide for each subject, and the self-assessment are presented using the NCS in conjunction with the teaching plan [3].

2.3 NCS Education Services

The expression education service was first used in nonprofit marketing [48]. One of the few misconceptions that we often hear when talking about lifelong learning marketing is that "lifelong learning institutions are just as much a school as they are schools, why do they introduce marketing concepts into the management of lifelong learning institutions?" to be [50]. Educators say that education is not included in the service sector, but among them, scholars studying the field of economics of education, management and economics argue that education is included in the service field.

In addition, education is classified as a service industry (Statistics Korea, 2014) in Korea's standard industry classification, and education is included in the service classification of GATT(General Agreement on Tariffs and Trade [7].

Many scholars' studies on education services have shown that educational services are defined as intangible services provided to people [7]. Educational services are also the aggregate of all human, material and organizational behaviors that are provided to the learners. Educational services are defined as the activities of a provider from the perspective of the institution and the bundle of benefits as all of the experience from the perspective of the learners [36].

Based on previous studies, it can be said that enhancing service quality in the NCS education service industry, as in other industries, is a differentiated strategy for acquiring competitive advantage.

3 Research Design

In this study, the research areas of Seo (2013) and Kim (2011) were compared with the evaluation area, items and indicators of NCSE accreditation. Through this, we classify the educational service quality of SERVQUAL model as educational factor, administrative factor, and facility factor [7, 14].

We derive research models and hypotheses based on previous studies to achieve the purpose of the study. In addition, we set up a research model to examine how the factors related to NCS education service quality affect perceived value and educational satisfaction. We also describe the operational definitions and measurement items of the used variables.

3.1 Research Model

In order to investigate the factors affecting the satisfaction of education on NCS education service quality, we set up a research model as shown in [Fig. 1].

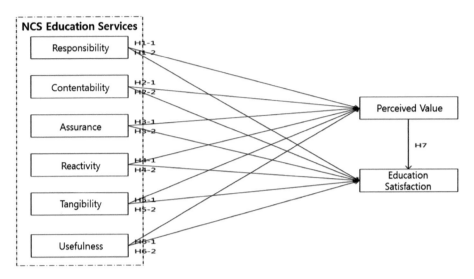

Fig. 1. Research model

3.2　Operational Definition of Variables

In this study, NCS education service quality factors were classified into responsibility (NCS curriculum), contentability(practice theory), assurance(teaching/lecturer), and reactivity(education administration).

Responsibility, contentability, assurance and reactivity explained in the theoretical background were reconstructed as quality factors of education service in this study.

Therefore, in order to verify the relationship between the variables and the research model, a total of 13 research hypotheses were set up and the research hypothesis was verified (Table 1).

Table 1. Operational definition of variables

Variable	Operational definition	Researcher
Responsibility (NCS Training course)	The extent to which the NCS education system is capable of providing and performing services that are accurate and reliable	PZB, 1996 [60] Rust and Oliver, 1994 [56]
Contentability (Practical Theory)	The contents of the education provided by the NCS education system and the contents of various educational programs	Hwang, 2005 [34] Kim, 2008 [6] Shim, 2009 [17]
Assurance (Professor/Lecturer)	The knowledge of the faculty/lecturer who provides the NCS education system, and the degree of ability to provide faith and stability	PZB, 1996 [60] Rust and Oliver, 1994 [56]

(continued)

Table 1. (*continued*)

Variable	Operational definition	Researcher
Reactivity (Education Administration)	The extent to which the NCS education system responds quickly to the needs of the target audience	PZB, 1996 [60] Rust and Oliver, 1994 [56]
Tangibility (Educational environment)	The degree to which physical facilities, tools that provide NCS education systems are evaluated	PZB, 1996 [60] Rust and Oliver, 1994 [56]
Usefulness (Employment support)	The degree of usefulness of the various expertise, career, and performance that the NCS education system provides to the target audience	InHwang, 2005 [34] Kim, 2008 [6] Soojin Shim, 2009 [17]
Perceived value	The size of the cost benefit from using the NCS education system	Zeithmal, 1988 [59] Woodruff, 1997 [57] Gardial, 1996 [46]
Education Satisfaction	Satisfaction with experience using NCS education system	Oliver, 1980 [53] Anderson, 1994 [37]

3.3 Hypothesis

(1) Hypotheses on NCS Education Service Quality and Perceived Value

Bolton and Drew suggested that service quality is formed by satisfaction and dissatisfaction based on service purchase and consumption experience, and that service quality may affect service value [40].

Dodds found that perceived service quality and perceived cost form perceived service value in "Study on the influence of price and store names" in relation to service quality evaluation, and perceived service value has the greatest influence on customer behavior Respectively [44].

Rust argues that perceived value is a determinant of purchasing and repurchasing, and perceived value is higher when the service quality is ideally evaluated because perceived value is highly correlated with quality or price [56].

Choi said that students who received excellent education service quality had significant influence on service value and increased student satisfaction [32].

Lee demonstrated that service quality such as physical environment, human environment, privacy have a significant influence on service value in the study on the effect of service quality of university dormitory on service value, student satisfaction and intention to reuse [26].

Based on the above results, it is assumed that there is a significant structural causality between NCS education service quality and perceived service value in the university (Table 2).

Table 2. NCS education hypotheses on service quality and perceived value

Hypothesis	
H1-1	Responsibility(NCS curriculum) will have a positive impact on perceived value
H2-1	Contentability(practice theory) will have a positive effect on perceived value
H3-1	Assurance(professor/lecturer) will have a positive effect on perceived value
H4-1	Reactivity (education administration) will have a positive effect on perceived value
H5-1	Tangibility (educational environment) will have a positive effect on perceived value
H6-1	Usefulness(employment support) will have a positive impact on perceived value

(2) Hypotheses on NCS Education Service Quality and Perceived Value

Although it is unclear which variable is a leading variable in the structural causality between service quality and educational satisfaction, it is emphasized that service quality is a leading variable of educational satisfaction, and many previous studies have proved this.

Woodside et al. (1989) presented the causal relationship between service quality, satisfaction, and intention for the first time [58]. In addition, Cronin and Taylor (1992) investigated the factors affecting service quality and identified the determinants of service quality and analyzed the relationship between satisfaction. They found that service quality had a significant effect on satisfaction, and that satisfaction had a significant effect on intention in the structural analysis of satisfaction and service quality and causality of intention [43].

Based on these previous studies, we conclude that there is a significant structural causality between the quality of education service and the satisfaction of education (Table 3).

Table 3. Hypothesis on NCS education service quality and educational satisfaction

Hypothesis	
H1-2	Responsibility(NCS curriculum) will have a positive impact on educational satisfaction
H2-2	The contentability of education (practical theory) will have a positive effect on educational satisfaction
H3-2	Assurance(professor/lecturer) will have a positive effect on the satisfaction of education
H4-2	Reactivity (educational administration) will have a positive effect on educational satisfaction
H5-2	Tangibility (education environment) will have a positive effect on educational satisfaction
H6-2	Usefulness(employment support) will have a positive (+) impact on educational satisfaction

Hypotheses on NCS Education Service Quality and Perceived Value

(3) Hypotheses on NCS Education Service Quality and Perceived Value

Hwang (2005) selected perceived values as factors that have an important effect on educational satisfaction. The results showed that the perceived value had a significant effect on educational satisfaction and that perceived value has a significant effect on educational satisfaction [34].

Oh (2011) studied the effect of perceived value on vocational high school students' educational satisfaction and confirmed that perceived value had a significant effect on educational satisfaction [22].

As mentioned earlier, based on previous studies, it is judged that there is a significant structural causal relationship between perceived value and educational satisfaction. In this study, the following hypothesis is set and verified (Table 4).

Table 4. Hypotheses on perceived value and educational satisfaction

Hypothesis	
H7	Perceived value will have a positive(+) effect on educational satisfaction.

4   Empirical Analysis

In this study, students and graduates who use NCS education service quality questionnaire were surveyed from March 17, 2017 to April 6, 2017 for 3 weeks. As a result, 285 questionnaires were used for the empirical analysis except for the missing value and unsatisfactory answer questionnaire.

4.1   Characteristics of Respondents

The demographic characteristics of the sample for empirical analysis are shown in Table 5.

Table 5. Characteristics of respondents

Division		Frequency (persons)	Ratio(%)
Gender	Male	113	39.1
	Female	176	60.9
Age	20's	289	100

4.2   Feasibility and Reliability Analysis

In this study, exploratory factor analysis (EFA) was conducted using the statistical program SPSS 22 to analyze the validity and reliability of the factors, and the validity and reliability of each factor were verified as shown in Table 6 below.

Table 6. Exploratory factor analysis

Configuration concept	Ingredient								Cronbach's α
	1	2	3	4	5	6	7	8	
a1	.199	.757	.152	.250	.251	.118	.200	.104	0.938
a2	.197	.827	.132	.140	.137	.204	.138	.176	
a3	.180	.784	.227	.188	.177	.159	.236	.160	
a4	.217	.773	.213	.165	.188	.199	.117	.198	
b2	.291	.288	.187	.229	.286	.193	.652	.123	0.888
b3	.246	.227	.185	.191	.167	.187	.743	.178	
b4	.252	.208	.237	.169	.222	.194	.693	.240	
c1	.826	.194	.161	.200	.167	.113	.105	.174	0.943
c2	.763	.211	.180	.200	.230	.137	.192	.198	
c3	.792	.177	.241	.192	.161	.157	.213	.160	
c4	.760	.219	.227	.157	.228	.149	.248	.113	
d1	.225	.151	.220	.747	.182	.194	.215	.114	0.928
d2	.281	.271	.386	.659	.193	.191	.062	.177	
d3	.231	.265	.249	.715	.181	.287	.148	.167	
d4	.190	.252	.371	.668	.159	.174	.251	.179	
e1	.204	.205	.760	.245	.135	.193	.086	.217	0.924
e2	.209	.169	.748	.197	.178	.274	.207	.154	
e3	.303	.170	.675	.288	.192	.204	.230	.195	
e4	.211	.227	.692	.314	.171	.201	.172	.112	
f1	.081	.251	.390	.160	.251	.668	.245	.161	0.936
f2	.203	.236	.370	.253	.272	.625	.238	.187	
f3	.240	.252	.234	.306	.201	.655	.182	.269	
f4	.223	.228	.224	.289	.242	.711	.143	.244	
g1	.294	.304	.194	.240	.373	.211	.252	.604	0.951
g2	.323	.246	.212	.225	.361	.238	.277	.594	
g3	.254	.264	.326	.181	.225	.325	.178	.666	
g4	.258	.293	.305	.208	.223	.280	.233	.632	
h1	.332	.308	.230	.236	.678	.258	.223	.228	0.975
h2	.280	.280	.219	.236	.692	.252	.222	.248	
h3	.303	.275	.196	.209	.699	.257	.248	.235	
h4	.311	.309	.273	.198	.661	.279	.252	.240	
Eigen Value	4.130	4.076	3.790	3.275	3.184	3.021	2.633	2.521	N/A
% of Variance	13.324	13.149	12.226	10.566	10.271	9.745	8.494	8.133	
Cumulative(%)	13.324	26.473	38.699	49.265	59.536	69.281	77.775	85.907	

a: reliability, b: contentability, c: assurance, d: reactivity, e: tangibility, f: usefulness, g: perceived value, h: satisfaction

4.3   Validation of Discrimination

If the square root of the mean variance extracted value for the measurement variable is larger than the correlation between the conceptual variables, it is judged that there is discriminant validity between the variables [45].

Table 7. Exploratory factor analysis

	A	B	C	D	E	F	G	H
A	0,762							
B	0,260	0,735						
C	0,156	0,300	0,723					
D	0,237	0,286	0,233	0,762				
E	0,180	0,284	0,235	0,479	0,766			
F	0,247	0,346	0,185	0,420	0,457	0,746		
G	0,298	0,428	0,311	0,335	0,357	0,498	0,745	
H	0,298	0,428	0,318	0,308	0,292	0,450	0,552	0,822

A: reliability, B: contentability, C: assurance, D: reactivity, E: tangibility, F: usefulness, G: perceived value, H: satisfaction

As shown in Table 7, since the square root value of the average variance extracted value of each variable is larger than the correlation coefficient, it is confirmed that there is no problem in the discriminant validity between the constituent variable concepts.

4.4 Hypothesis Test Summary

In the hypothesis that responsibility (NCS Curriculum), contentability (Practice Theory), assurance (Education/Instructor), reactivity (Educational Administration), tangibility (Educational Environment), usefulness (Employment Support) as an independent variable in service quality affects perceived value, only four factors have a positive(+) effect on perceived value: responsibility(NCS curriculum), contentability(practical theory), assurance (education/lecturer), and usefulness (employment support). Among the factors, the factors that have the greatest influence on the perceived value were usability, followed by contentability, assurance, and responsibility (Fig. 2).

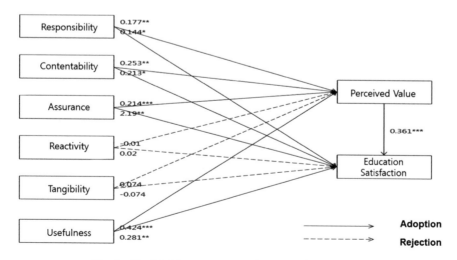

Fig. 2. The Model to measure confirmatory factor analysis

The independent variables of NCS education service quality affecting educational satisfaction were responsibility(NCS curriculum), contentability(practical theory), assurance (education/lecturer), usefulness (employment support), and reactivity (education administration) and tangibility (education environment) did not affect the satisfaction of education positively. The independent variables which had the greatest influence on the satisfaction of the education were usefulness (job support), followed by contentability(practical theory), assurance (education/lecturer), and responsibility (NCS curriculum).

Finally, perceived value has a positive effect on the degree of educational satisfaction recommendation intention.

5 Conclusion

According to the results of the analysis, factors that can influence "perceived value" through NCS are curriculum, educational theory, ability of instructor/lecturer, and actual employment support.

The implication of this study is that the urgent thing in the NCS education system is "efficient and effective management of job skills required in the industrial field".

Therefore, it is necessary to involve practitioners in each industry in designing NCS curriculum. Now, the design of NCS curriculum is mainly carried out by educational experts. As a result, the design of the NCS curriculum focused mainly on the contents of education. However, as suggested in this study, it is necessary to involve practitioners in each industry from the time of designing the NCS curriculum. If it does, it will be able to achieve the NCS's primary purpose of cultivating job skills during study.

References

1. Mansu, K., Park, S.-K.: Assessing the effects of service quality on student satisfaction, reputation and loyalty: the case of university education. Educ. Adm. Korean Soc. Study Educ. Adm. **29**(1), 153–174 (2011)
2. Gu, E.: Participants' Perception on the NCS-Based Inservice Vocational Training Program. Major in Curriculum & Instruction, The Graduate School of Education, Ewha Womans University (2016)
3. The Ministry of Education: NCS Education Status and Actual Condition. National Research Foundation of Korea (2015)
4. Su, K.K.: (AMOS 18.0) Structural Equation Modeling. Hannarae Publishing Co. (2010)
5. Kim, D.-H., Kim, D.-U.: AMOS A to Z (Analysis of Structural Equation Model Based on the Writing Process). Hakhyunsa (2009)
6. Kim, W.-I.: (An) analysis on relationship between components of educational service quality in general high school and satisfaction & loyalty level with school education. Major in Educational Managementand Lifelong Education, Graduate School of Education, Yonsei University (2008)

7. Kim, J.: An empirical study on effect of education service quality which influence re-use and customer satisfaction - focusing on E life-long education center. Korean Soc. Qual. Manag. **39**(1), 155–166 (2011). (12 pages)
8. Rha, H.-M., Shim, J.-H.: Development of S-OJT training program for a Software Company based on National Competency Standards (NCS): a case of firm A. Korean Ind. Educ. J. **39**(2), 81–100 (2014). (20 pages)
9. Bae, B.-L.: Structural equation modeling whit AMOS 21-Principle and Practice. Chungram (2014)
10. Bae, I., Choi, J., Kang, M., Lim, S.: The effects of service quality of education on service commitment - focused on life insurance planners. Korean Soc. Qual. Manag. **41**(1), 79–94 (2013). (16 pages)
11. Park, K.-H.: The influence of perceived educational service quality on students satisfaction and relationship marketing of students majoring in dance. Korean J. Phys. Educ. **48**(6), 463–473 (2009). (11 pages)
12. Park, J.W., Lee, J.S.: The relationship among service quality, customer satisfaction, image, and service loyalty in fast food service. Korean Mark. Manag. Assoc. **7**(1), 3–69 (2002). (67 pages)
13. Park, E., Hong, S.H., Oh, S.: A study on university students' perception of National Competency Standards (NCS). Korea Acad. Soc. Tour. Manag. **30**(4), 29–52 (2015). (24 pages)
14. Seo, D.-G.: A Study on the Effect of Education Service Quality and Institutes Satisfaction of Academy Credit Bank System for Lifelong Education on Perceived Academic achievement, Re-enrollment Intention and Word-of-Mouth Intention - Focused on Cosmetology-Majored Students, Graduate School of Education, Chosun University (2013)
15. Sung, H.S., Han, S.-L.: A study on perceived quality affecting the service personal value - focusing on the moderating effect of Need For Cognition (NFC) and Need For Touch (NFT). Korean Soc. Consum. Stud. **20**(3), 163–188 (2009)
16. Song, J.: SPSS/AMOS statistical analysis method for writing paper, 21CSA (2012)
17. Shim, S.: (An) analysis of relationship between components of educational service quality of middle school students and satisfaction. Major in Educational Administration and Life-long Education, Graduate School of Education, Yonsei University (2009)
18. Shim, W.-G.: Satisfaction, customer orientation and service delivery level in university. J. Creat. Innov. (JCI), Satisf. Cust. Orientat. Serv. Deliv. Level Univ.
19. Woo-Geuk, S., Choi, H.: A study on the structural relationships between internal marketing and organizational performance in university. Soc. Sci. Res. Inst. **38**(2), 93–124 (2014). (32 pages)
20. Woo-Geuk, S., Choi, H.: The effects of educational service quality on service value, customer satisfaction, voluntary behavioral intention: focused on the moderating effects of value congruence. Social Science Research Institute **54**(2), 181–215 (2015). (35 pages)
21. Oh, M.: A Study on the Reaction Evaluation of Job Analysis-based Curriculum: Focusing on the NCS-Based Curriculum of Community Colleges. Major in Education Consulting, Graduate School of Education, DongDuk Womens University (2015)
22. Oh, S., Song, J.-H., Shim, K.: The effects of educational service orientation and service value on student satisfaction and university image: the strengthening plan of competitiveness in university educational service. J. Mark. Stud. **19**(4), 23–42 (2011). (20 pages)
23. Woo, J.-P.: Structural Equation Model Concept and Understanding. Hannarae Publishing Co. (2012)
24. Cheol, L.K., Lee, M.: Study of cause and effect model among variables related to service quality for enhancing productivity of university education. Korea Product. Assoc. **21**(2), 1–35 (2007)

25. Lee, K.-H., Hyun, S.: The effect of brand experience on brand attachment, word-of-mouth intention, and revisit intention in the context of convention center: focusing on moderating role of customer involvement. Korean Hosp. Tour. Acad. **23**(4), 통권 80 호, 113–127 (2014). (15 pages)
26. Lee, S., Park, J.-M.: Influence of the service quality on the service value, student satisfaction and reuse intention in the university dormitory. Korean Ind. Econ. Assoc. **23**(6), 3017–3033 (2010). (17 pages)
27. Lee, J.-H.: The moderating effects of self-participation regarding the impact of education service quality on student satisfaction - focusing on the major of food service and culinary arts. Culin. Soc. Korea **18**(1), 246–258 (2012). (13 pages)
28. Lee, H., Im, J.H.: Research methodology for writing social science articles using SPSS. Jyphyunjae (2014)
29. Lee, P.-J., Kim, Y.-R., Jeong, H.-J.: An empirical study on measurement and improvement for the service of education and administration in the university. Korean Soc. Comput. Inf. **13**(3), 197–210 (2008). (14 pages)
30. Young, L.H.: Research Methodology. Chungram (2010)
31. Chae, S.-I.: Social Science Research Methodology. BNM Books (2005)
32. Choi, D.-C., Lee, K.-o.: A study on building a marketing model for university education services. J. Glob. Acad. Mark. Sci. **6**, 339–366 (2000)
33. Choi, S., Cho, Y.: A study of the effect of LOHAS traits on customer participation, customer satisfaction, customer loyalty for Korean traditional pastry customer. Foodserv. Manag. Soc. Korea **13**(2), 169–189 (2010). (21 pages)
34. Hwang, I.-H.: (A) study on the measurement of educational service quality and structural causality of student satisfaction, Department of Business Administration, Graduate School, Gyeongsang National University (2005)
35. Jun, H.: Heo Jun's Easy-To-Follow AMOS Structure Equation Model. Hannarae Publishing Co. (2013)
36. Hong, G.S.: A Study on Each Level of Educational Service Quality on Customer's satisfaction and Loyalty - Focused on Quality of College Educational Service, Department of Business Administration Graduate School of Myong Ji University (2005)
37. Anderson, R.A.: Stress Effects on Chromium Nutrition of Humans and Farm Animals, pp. 267–274. Nothingham University Press, Nothingam (1994)
38. Bentler, P.M., Bonett, D.G.: Significance tests and goodness of fit in the analysis of covariance structures. Psychol. Bull. **88**, 588–606 (1980)
39. Bentler, P.M.: Comparative fit indexes in structural models. Psychol. Bull. **107**(2), 238–246 (1990)
40. Bolton, R.N., Drew, J.H.: A Longitudinal analysis of the impact of service changes on customer attitudes. J. Mark. **55**(1), 1–10 (1991)
41. Browne, M.W., Michael, W., Cudeck, R.: Alternative ways of assessing model fit. Sociol. Methods Res. **21**, 230–258 (1992)
42. Carmines, E.G., McIver, J.P.: Analyzing models with unobserved variables: analysis of covariance structures. In: Bohrnstedt, G.W., Borgatta, E.F. (eds.) Socia Measurement: Current Issues, pp. 65–115. Sage Publications, Beverly Hills (1981)
43. Cronin Jr., J.J., Taylor, S.A.: Measuring service quality: a reexamination and extension. J. Mark. **56**, 55–68 (1992)
44. Dodds, W.K.: Factors associated with dominance of the filamentous green alga Cladophora glomerata. Water Res. **25**(11), 1325–1332 (1991)
45. Fornell, C., Larcker, D.F.: Evaluating structural equation models with unobservable variables and measurement error. J. Mark. Res. **18**(1), 1981 (1981)

46. Gardial, S., Woodruff, R.B.: Know your customer: new approaches to understanding customer value and satisfaction. Wiley, New York (1996)
47. Hair Jr., J.F., Black, W.C., Babin, B.J., Anderson, R.E.: Multivariate Data Analysis, 7th edn. Pearson Hall, Upper Saddle River (2010)
48. Joreskog, K.G., Sorbom, D.: LISREL VI: analysis of linear structural relationships by the method of maximum likelihood. National Educational Resources, Chicago(1984)
49. Kotler, P.: Overview of political candidate marketing. NA-Advances in Consumer Research, vol. 2 (1975)
50. Kotler, P., Fox, K.: Strategic Marketing for Educational Institutions, 2nd edn. Prentice-Hall, New York (1995)
51. Mulaik, S.A., James, L.R., Van Alstine, J., Bennet, N., Lind, S., Stilwell, C.D.: Evaluation of goodness-of-fit indices for structural equation models. Psychol. Bull. **105**(3), 430–435 (1989)
52. Muthén, B., Kaplan, D.: A comparison of some methodologies for the factor analysis of non-normal Likert variables. Br. J. Math. Stat. Psychol. **38**, 171–189 (1985)
53. Oliver, R.L.: A cognitive model of the antecedents and consequences of satisfaction decisions. J. Mark. Res. **17**, 460–469 (1980)
54. Parasuraman, A., Zeithaml, V.A., Berry, L.L.: A conceptual model of service quality and its implications for future research. J. Mark. **49**, 41–50 (1985)
55. Parasuraman, A., Zeithaml, V.A., Berry, L.L.: Servqual: a multiple-item scale for measuring consumer perc. J. Retail. **64**(1), 12 (1988)
56. Rust, R.T., Oliver, R.W.: The death of advertising. J. Advert. **23**(4), 71–77 (1994)
57. Woodruff, R.B.: Customer value: the next source for competitive advantage. J. Acad. Mark. Sci. **25**, 139 (1997)
58. Woodside, A.G., Frey, L.L., Daly, R.T.: Linking sort/ice anlity, customer satisfaction, and behavioral intention. J. Health Care Mark. **9**(4), 5–17 (1989)
59. Zeithaml, V.A.: Customer perceptions of price, quality and value: a menas end model and synthesis of evidence. J. Mark. **52**(1), 2–22 (1988)
60. Zeithaml, V.A., Berry, L.L., Parasuraman, A.: The behavioral consequences of service quality. J. Mark. **60**, 31–46 (1996)

Model-Driven Development of Mobile Applications Allowing Role-Driven Variants

Veiga Yamodo Marland[1] and Haeng-Kon Kim[2(✉)]

[1] Computer Information Engineering, Daegu Catholic University,
Hayang-up, Gyeongsan-si, Gyeongsangbuk-do, South Korea
`yams24@naver.com`
[2] Daegu Catholic University,
Hayang-up, Gyeongsan-si, Gyeongsangbuk-do, South Korea
`hangkon@cu.ac.kr`

Abstract. This research aims to simplify the creation of applications for mobile platforms by developing a high-level and platform independent model of an application, and automatically transforming this high-level model to platform specific code. The research method is a combination of the model-driven development (MDD) approach in software development and application of techniques in the field of human-computer interaction (HCI) particularly on user centered system design. This research involves developing a graphical modeling language which is specific to mobile applications, and coming up with a generic algorithm for the conversion of this graphical model into code. The main focus of the research however, will be on the design of the graphical model, and the interaction techniques which will allow non-expert people 1 to create specialized mobile applications with ease. Key research questions that need to be answered are: Which level of abstraction and modeling constructs are adequate for non-expert people to create specialized mobile applications? How can technical details such as device limitations, transmission of information to other devices or to the network, etc., be abstracted from the non-expert user as they model the application so that they can focus more on the design and logic of their application, but at the same time take advantage of advance capabilities of mobile devices (GPS, Bluetooth, Wi-Fi, etc.)? How can the design of interaction with other devices that connects to mobile applications become simpler? Do non-experts want the modeler interface to look like the actual application or do some high level form just like in UML would be enough?

Keywords: Model-Driven Development · Mobile application models
Human-computer interaction

1 Introduction

The use of mobile applications has become an indispensable part of daily life.

This has already and will lead to rapidly increasing numbers of applications and users that make the development of mobile applications to one of the most promising fields in software engineering. Mobile application development faces.

© Springer Nature Switzerland AG 2019
R. Lee (Ed.): ACIT 2018, SCI 788, pp. 14–26, 2019.
https://doi.org/10.1007/978-3-319-98370-7_2

Several specific challenges that come on top of commonplace software production problems. Popular platforms differ widely in hardware and software characteristics and typically show short life and innovation cycles with considerable changes. The market often requires that apps must be available for several platforms which makes a very time and cost-intensive multiple platform development a necessity. Available solutions try to circumvent this problem by using web-based approaches, often struggling with restricted access to the technical equipment (e.g. sensors) of the mobile device and making less efficient use of the device compared to native apps. Furthermore, web-based solutions require an app to stay on-line more or less permanently which may cause considerable costs and usability restrictions. Model-driven development (MDD) can help to improve this situation by a faster and easier development process of native apps and easier adaptability to new features of underlying platforms. Mobile apps are modeled in a modeling language focusing on main system aspects being enough to automatically generate platform-specific code. Hence, MDD allows to develop applications on a higher abstraction level, neglecting technical details.

Furthermore, code quality may improve since code can be generated according to standards. In addition, the validation of app models w.r.t. system requirements may be facilitated by deducing formal models on a similar abstraction level.

The heart and soul of model-driven development is the domain-specific modeling language (DSML). Our developed DSML covers three aspects of mobile apps: its data entities and relations, its behavior including data management, access of sensors, use of other apps, etc., and its user interface. The design of our modeling language follows the credo: "Model as abstract as possible and as concrete as needed." This means the following: Standard solutions are modeled very abstractly while more specific solutions are modeled in more details. E.g. data management by the usual CRUD functionality (using create, read, update, and delete operations) may be modeled by one predefined model element type while application-specific behavior is specified on the level of usual control structures.

To efficiently work with the DSML, we provide an Eclipse-based tool environment consisting of a graphical editor with three different views for data, behavior and user interface models as well as two code generators to Android and iOS. Before developing these two code generators, we studied the design principles of mobile apps in Android and iOS and found a lot of commonalities.

The idea behind this work is to develop apps for different platforms as similar as adequate. It is possible to use the same overall architecture independent of the chosen platform. We demonstrate the potentials and limits of our MDD approach at a small example.

Considering our approach to MDD of mobile apps, the following issues are still open: Due to spatial movement, mobile apps offer new possibilities to interact with the environment: They support an increasing variety of sensors such as cameras, global positioning system (GPS), compass, etc. which can be advantageously used to position a mobile device and to inform the user about its position. These capabilities are especially interesting for augmented reality AR) and navigation in space. Mobile applications claim to operate reliably using spatial movement, however, developers have to deal with the effects of hanging environmental contexts. One of the most important contexts is the connectivity of mobile devices. Since mobile applications are increasingly used s front-ends of transaction systems, they have to be designed for being able to deal with intentional or accidental loss of connection. In order to support higher mobility - in the sense that operations may execute across the boundaries of changing network states - problems and requirements for context-aware architectures of mobile applications are considered. We discuss how these issues may be supported by model-driven development as well.

A faster and easier development of mobile apps is nothing if their quality cannot be assured. We discuss what software quality means for mobile apps and what kinds of quality assurance techniques may be promising.

1.1   Agile Development of Mobile

The agile methodology for mobile application development aims to provide an alternative to traditional project management methods such as cascade methodology. This approach allows requirements and solutions to evolve through the combined efforts of the development team and the customer. It promotes adaptive planning, scalable development, early delivery and continuous improvement. This iterative and flexible approach can be used in complex projects where customer requirements change frequently. A large project can be divided into smaller parts and an agile methodology can be applied to each of these small parts. As this method requires a high commitment from the client to take into account the client's requirements at each step and his comments after each step, it can be used in projects where the client commits to engaging in an interactive time communication in time. Now, when it comes to understanding the role of agile methodology for mobile application development, one must take into account the fact that mobile application developers are creating an application with a small screen, less memory and less processing speed. In addition, with the number of mobile devices with different operating systems and different operators, creating a mobile application for all devices becomes a difficult task for developers. Apart from this, developers also face short life cycle development issues, limited hardware, fast-changing technology, and changing user demands as technology changes. In addition, developers also need to create an application that can be updated easily, can be downloaded easily and has an excellent UX design. In the end, having seen the endless number of requirements that a mobile application development company must fulfill in order to build a remarkable application, the first question that comes to mind is that how companies can then render user-friendly, high-quality applications? [1].

2   The Mobile Application Domain

2.1   Why Is Architecture Important?

Mobile apps are developed for a variety of purposes from simple entertainment to serious business applications. We are moving towards a kind of commercial application where basic generic building blocks are provided for a selected domain. These building blocks can be used and refined by domain experts to customize them according to their specific needs. The fully customized application is then ready to be used by the end users. Consider concrete scenarios as they occur in our collaboration with advanced, the industrial partner of our project: key2guide is a multimedia guide that can be configured without programming. Its typical application lies in the context of tourism where visitors are guided through places of interest, egg. A museum, an exhibition, a city or a region. Objects of interest are listed and explained by enriched information. In addition, objects can be categorized and commissioned in additional structures, that is, tours that guide visitors through an exhibition. As the reader can expect, such an application is rather data-driven. These data usually change frequently over time. As a result, a typical requirement is to provide the opportunity for domain experts (for example, museum administrators or managers) to update the data regularly. In addition, moving around a region could lead to restricted Internet connections. Therefore, Web applications would not be preferred solutions. On the other hand, applications usually have to work offline, but can download new vendor information from time to time. A second advanced product, called key2operate, is used to define manual business processes with support for mobile devices to integrate into a holistic production process. For example. In order to inspect the machines of a production plant, the worker receives a list of inspection requests that must be executed sequentially. Such an execution could include the collection of critical data. Machines can be identified by scanning bar codes or reading RFID chips. The control values can be entered manually by the worker. In addition, the start and end times of the execution can be taken. After completing an inspection request, the application must display the next request to be executed and direct the worker to the corresponding machine online. Again, an application is required to be configured by the users being the production managers defining here their intended business process. As production processes have become very flexible nowadays, manual processes with support for mobile devices must also be continuously adaptable. Key2operate allows such process adaptations without deploying it again. However, the process definitions are quite simple because they support simple data. Structures only. Both applications work with a Web-based backend content management system to maintain the configurations available to end users. In summary: We are moving towards mobile business model driven development that supports the configuration of user-specific variants. In this scenario, there are usually several types of users, e.g. providers that provide custom content and end users

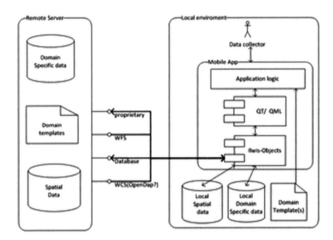

Fig. 1. Mobile app environment

who use a configured application with all the information provided. Of course, provider groups and end users can be more elaborately structured, so that different roles are defined. For example, a tourist guide for a city may cover sites in the city as well as several museums and exhibitions. Guidance information is usually provided by multiple vendors with different roles. The city's tourism officials are allowed to change the information on the sites in the city-only category, for example. The administrators of the history museum can edit all the information about the objects in their museum. Role-specific application variants need to be developed [2] (Fig. 1).

2.2   Mobile App Requirements

A Business Application Specification Document (PRD), also known as a product specification document, is the foundation for a product that describes business logic, lists technical specifications, and directs the team from the initial design stage to the final sprint stage. Product teams use the PRD to guide the development of mobile applications and understand what they need to build their products. Product Requirements the purpose of the document is to create an attractive product. In other words, you need to collect a lot of research on competencies, users, skills, and team competencies. Quality products start with the need to try to answer them. The product requirements document helps teams understand the product needs and how the product is used. While there are many ways to organize this document, we have found it useful to include both operating requirements and technical/product requirements as well as some considerations that will help engineers prepare to sell your product. Here's how to create a requirements document for an effective mobile application [3] (Fig. 2).

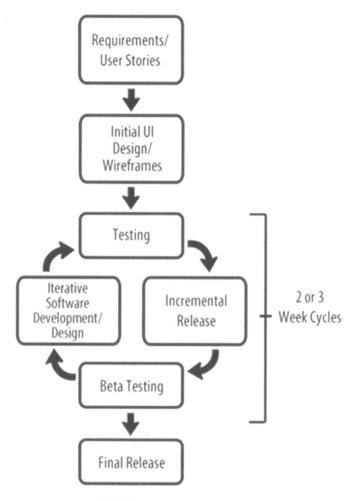

Fig. 2. Developing mobile app

3   MDD-Infrastructure for Mobile

A model driven development (MDD) infrastructure has great potential for accelerating the development of software applications. While simply modeling application-specific data structures, processes, and layouts, executable software systems can be generated. As a result, MDD does not focus on technical details, but elevates software development to a higher level of abstraction. The heart and soul of MDD is the domain-specific modeling language. It comes with a tool environment consisting of textual or visual model editors and code generators appropriate for the desired target platforms (for example, Android and iOS). For the development of our MDD infrastructure, we have chosen an agile bottom-up process [VSRT15], starting with the analysis of a domain and the identification of mobile application features, the extraction of templates from re-implemented prototypes and iterative language extension [4] (Fig. 3).

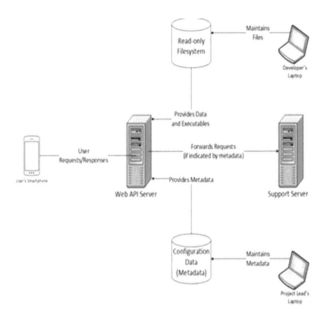

Fig. 3. Mobile platform and mobile app

3.1   Overview of Development Process

To solve the problem mentioned above, we propose a UML-based MDD method using a specific smartphone function model and an original GUI constructor independent of any specific operating system. Figure 4 shows an overview of our method.

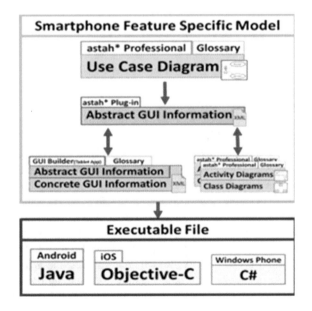

Fig. 4. Overview of our method

3.2   UML Models

Associated actor, such as an available external user or application or hardware component. A developer can decide on a root use case by linking the other use cases by using extension or inclusion relationships. The root use case represents a startup scenario for the application. At the end of the operation, each use case is defined by an activity diagram and the relation is expressed by calling a sub-activity corresponding to the other use case in the activity diagram.

An activity diagram expresses a series of processing actions with related data. The background action is distinguished from the foreground action by the use of a partition. An object node is used to designate the binding of external applications or hardware. A user partition includes user actions with input data while the Interaction partition includes actions with output data through a user interface. Figure 5 shows a class structure of the system partition specified by the glossary state. In this model, the general components of an activity diagram are specified by the features of the smartphone. A glossary class is displayed in red and a developer can design smartphone features using this class [5].

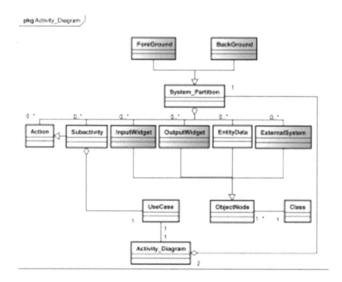

Fig. 5. System partition in activity diagram

4   The Mobile Applications Domain

Mobile apps are developed for very diverse purposes ranging from mere entertainment to serious business applications. We are heading towards a kind of business app where basic generic building blocks are provided for a selected domain.

These building blocks can be used and refined by domain experts to customize them according to their specific needs. The fully customized app is then ready to be used by end users. Let's consider concrete scenarios as they occur in our collaboration with advanced, the industry partner of our project: key2guide is a multimedia guide that can be configured without programming. Its typical application lies in the context of tourism where visitors are guided through places of interest, e.g. a museum, an exhibition, a town or a region. Objects of interest (e.g. paintings, crafts and sculptures presented in a museum) are listed and explained by enriched information. Furthermore objects may be categorized and ordered in additional structures, i.e. tours that guide visitors through an exhibition. As the reader might expect, such an app is pretty data-oriented [6]. This data usually changes frequently over time. In consequence, a typical requirement is to offer a possibility that domain experts (e.g. museum administrators or tourism managers) can refresh data regularly. Moreover, moving around in a region might lead to restricted Internet connections. Hence, web apps would not be preferred solutions. In contrast, apps shall typically run off-line but can download new provider information from time to time. A second product by advanced, called key2operate, allows to define manual business processes with mobile device support to be integrated into a holistic production process. E.g. in order to inspect machines of a production plant, the worker gets a list of inspection requests that has to be executed sequentially [7]. Such an execution might include the collection of critical data (e.g. pressure or temperature). Machines can be identified by scanning bar codes or reading RFID chips. Control values might be entered manually by the worker. Moreover start and end times of the execution may be taken. After finishing an inspection request, the app shall display the next request to be executed and direct the worker to the corresponding machine in line. Again, an app is required that may be configured by users being production managers here defining their intended business processes. As production processes have become very flexible nowadays, manual processes with mobile device support also have to be continuously adaptable. Key2operate allows such process adaptations without newly deploying it. However, process definitions are pretty simple since they support simple data structures only. Both apps work with a web-based backend content management system to maintain configurations that are available for end users to summarize: We are heading towards model-driven development of mobile business apps that support the configuration of user-specific variants. In this scenario, there are typically several kinds of users, e.g. providers who provide custom content, and end users consuming a configured app with all provided information. Of course, the groups of providers and end users may be structured more elaborately such that different roles are defined. For example, a tourist guide for a town may cover sights in the town as well as several museums and Exhibitions. The guiding information is typically given by several providers with different roles. Tourism managers of the town are allowed to edit information about sights in the category town only, while e.g. administrators of the history museum may edit all the information about objects in their museum. Role specific app variants shall be developed [8].

4.1   Language Design

The core of an infrastructure for model-driven development is the modeling language. In the following, we first present the main design decisions that guided us to our modeling language for mobile applications. Thereafter, we present the defining meta-model including all main well-formedness rules restricting allowed model structures. To illustrate the language, we show selected parts of a simple phone book app model. Finally, the presented modeling language is discussed along design guidelines for domain-specific languages (Fig. 6).

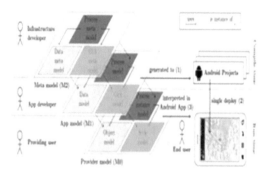

Fig. 6. Modeling approach

4.2   Design Decisions

Due to our domain analysis, we want to support the generation of mobile apps that can be flexibly configured by providing users. This main requirement is reflected in our modeling approach by distinguishing two kinds of models: app models specifying all potential facilities of apps, and provider models defining the actual apps. In the following Fig. 7, this general modeling approach is illustrated. While app models are used to generate Android projects (1) being deployed afterwards (2), provider models are interpreted by generated Android apps (3), can be used without redeploying an app. Instance models can be carried out in two ways: usually this will be done at runtime, because the instance model does not exist at build time, alternatively it can be done at build time, by adding the instance model to the resources of the generated android projects. The general approach to the modeling language is component-based: An app model consists of a data model defining the underlying class structure, a GUI model containing the definition of pages and style settings for the graphical user interface, and a process model which defines the behavior facilities of an app in form of processes and tasks. Data and GUI models are independent of each other, but the process model depends on them [9]. A provider model contains an object model defining an object structure as instance of the class structure in the data model, a style model defining explicit styles and pages for individual graphical user interfaces, and a process instance model selecting interesting processes and providing them with actual arguments to specify the actual behavior of the intended app. Similarly to the app model, object and style models are independent of each other but used by the process instance model.

(a) Main Menu with *Manage Persons* Process (CRUD functionality)

(b) *Persons Location* Process   (c) *Call Person* Process

Fig. 7. Screen shots of phone book app

5 Modeling Mobile Applications

The core of an MDD infrastructure is the domain-specific modeling language. It is used to model the specific aspects of applications in a so-called app model. This model is platform-independent and the input to available code generators, here, Android and iOS. The generator results are platform-specific software projects containing runnable apps. An app model consists of three sub-models: the data model, the process model to describe the app behavior, and the graphical user interface (GUI) model. In the following, we focus on the main language features. A detailed language definition can be found in. The data model contains the typical elements of object-oriented data modeling like classes, attributes, aggregations, associations, etc. It is not only used to generate the underlying data access objects (DAOs) but also determines the design of the user interface that is concerned with data input and output. The GUI model defines the graphical user interface of an app. It contains pages, style settings, and menus. The page type indicates the purpose of the page. Thus, app modelers describe the required user interface just by selecting a page type and the code generator deduces the detailed structure of the interface by consulting the data model. An app model consists of three sub-models: the data model, the process model to describe the app behavior, and the graphical user interface (GUI) model. In the following, we focus on the main language

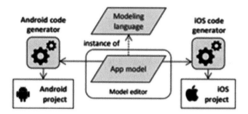

Fig. 8. Model-driven development of mobile apps

features. A detailed language definition can be found in. The data model contains the typical elements of object oriented data modeling like classes, attributes, aggregations, associations, etc. It is not only used to generate the underlying data access objects (DAOs) but also determines the design of the user interface that is concerned with data input and output. The GUI model defines the graphical user interface of an app. It contains pages, style settings, and menus. The page type indicates the purpose of the page. Thus, app modelers describe the required user interface just by selecting a page type and the code generator deduces the detailed structure of the interface by consulting the data model (Fig. 8).

6 Conclusion and Future Work

Mobile application model development is a promising approach to coping quickly Emerging technological development for several mobile platforms as well as time-to-market with the support of several, if not all, existing platforms. In this article, a modeling language for mobile applications is presented that allows to model mobile applications as abstract as possible and as concrete as necessary. Different user roles are not combined in one application, but lead to several application variants that can be configured after code generation, that is, content providing users, for end users. The domain under consideration is business applications driven by data or events, such as tour and conference guides, as well as manual sub processes in production processes. A selection of applications examples under development with our MDD-tool environment, can be found at. Future work will focus on other platforms, a code generator for iOS is currently being developed, and language extensions for flexible sensor manipulation and augmented reality. In addition, generated applications must be evaluated against. Software quality criteria, especially usability, data management, energy efficiency and secure Class library.

Acknowledgement. This Research was supported by the MSIP (Ministry of Science, ICT and Future Planning), Korea, under the ITRC (Information Technology Research Center) support program (IITP-2018-2013-0-0087) supervised by the IITP (Institute for Information & Communications Technology Promotion).

References

1. Starov, O., Vilkomir, S., Kharchenko, V.: Cloud testing for mobile software systems concept and prototyping. http://www.cyber-investigator.org/wpntent/uploads/2016/02/ICSOFT-EA_2013_7_CR.pdf
2. Knoop, J., Zdun (Hrsg.), U.: Software Engineering 2016. Lecture Notes in Informatics (LNI), p. 99. Gesellschaft fur Informatik, Bonn (2016)
3. Nuccini, H., Di Francesco, A., Esposito, P.: Software testing of mobile applications: challenges and future research directions. In: 7th International Workshop on Automation of Software Test, pp. 29–35. IEEE Press, New Jersey (2012)
4. Wasserman, A.I.: Software engineering issues for mobile application development. In: The FSE/SDP Workshop on Future of Software Engineering Research, FoSER 2010, pp. 397–400 (2010)
5. Berardinelli, L., Cortellessa, V., Di Marco, A.: Performance modeling and analysis of context-aware mobile software systems. In: Rosenblum, D.S., Taentzer, G. (eds.) FASE 2010. LNCS, vol. 6013, pp. 353–367. Springer, Heidelberg (2010). https://doi.org/10.1007/978-3-642-12029-9_25
6. Vaupel S., Taentzer G., Harries J.P., Stroh R., Gerlach R., Guckert M.: Model-driven development of mobile applications allowing role-driven variants. Philipps-Universität Marburg, Germany
7. Balagtas-Fernandez, F.T., Hussmann, H.: Model-Driven Development of Mobile Applications Department of Computer Science, University of Munich. http://www.medien.ifi.lmu.de
8. Knoop, J., Zdun (Hrsg.), U.: Model-driven development of platform-independent mobile applications supporting role-based app variability. In: Software Engineering 2016, Lecture Notes in Informatics (LNI). Gesellschaft fur Informatik, Bonn (2016)
9. Model-driven engineering languages and systems. In: Proceedings, 17th International Conference, MODELS 2014, Valencia, Spain, 28 September–3 October 2014

E-Learning Adaptation and Mobile Learning for Education

Mechelle Grace Zaragoza[1], Haeng-Kon Kim[1(✉)], and Ha Jin Hwang[2]

[1] School of Information Technology,
Daegu Catholic University, Gyeongsan, Korea
`mechellezaragoza@gmail.com, hangkon@cu.ac.kr`
[2] Sunway University Business School, Sunway University,
Subang Jaya, Malaysia
`hjwang@sunway.edu.my`

Abstract. Mobile learning through the use of mobile wireless technology allows anyone to access information and learning materials from anywhere around the globe. Mobile learning, through the use of mobile technology, will allow the users to access learning materials and information from anywhere, at any time at any cost. With mobile learning, students will be empowered because they can learn when and where they want to. They can use wireless mobile technology for formal and informal learning, where they can access personalized and additional learning materials from the Internet or from the host organization. This document is timely because there is significant growth in the use of mobile technology by people around the world, especially in developing countries. This study shows how mobile learning can transform the delivery of education and training.

Keywords: Advance learning · Educational technology

1 Introduction

Mobile technology can be used to provide instructions and information to these remote areas without people leaving their geographical area. One of the main advantages of using mobile wireless technology is to reach people who live in remote places where there are no schools, teachers or libraries. This will benefit communities in such places since students and workers will not have to leave their families and jobs to travel to a different place to learn or access information. At the same time, business owners, agricultural workers and other industries can access information to increase productivity and improve the quality of their products. People who live in remote communities will have access to health information to improve their health, which will improve the quality of life. Finally, because remote access using wireless mobile technology reduces the need to travel, its use can reduce the carbon footprint of humanity on Earth to help maintain a cleaner environment.

Educational technology has combined these approaches and has dramatically accelerated the future trajectory of education. This is an exciting time to get involved in the education sector, whatever your role. The revolution in education not only affects

R. Lee (Ed.): ACIT 2018, SCI 788, pp. 27–36, 2019.
https://doi.org/10.1007/978-3-319-98370-7_3

the way students learn: this paradigm shift in particular has a significant impact on it. This affects the way teachers teach, the way schools are structured, the barriers between school and family life, and perhaps at its deepest level - affects the trajectory of the whole future of life.

The global future of humanity in these modern and changing times is uncertain, unstable and dynamic. For future generations to adapt to this uncertainty and create sustainability, it is vital that the way we teach them to do so can also be adapted with equal dynamism. This is not the case of the "old" educational paradigms.

Technology has changed our world in an unimaginable way. Mobile devices permeate our daily lives, providing unprecedented access to communication and information. Looking to the next decade and beyond, it seems clear that the future of mobile learning lies in a world where technology is more accessible, affordable and connected than it is today. However, technology alone, regardless of its ubiquity and usefulness, will not determine whether mobile learning benefits a large number of people. Designing effective mobile learning interventions requires a holistic understanding of how technology intersects with social, cultural and, increasingly, commercial factors. Technology itself is undeniably important, but just as important, if not more, is the way people use and see technology, a point that has been largely neglected. It is not because mobile devices can, for example, help improve women's literacy skills in low-income communities, but that these devices will be used for this purpose. Worldwide, women are much less likely than men to own and use mobile devices, and in many communities women are not encouraged to use mobile technology for any purpose. Mobile devices are often banned in schools and other schools, despite the considerable and often well-established potential to improve learning. Such prohibitions project a view that mobile devices are antithetical to learning, and this perspective, regardless of its factual validity, influences how people interact with technology. Over the next 15 years, the implementation of mobile learning projects and the pedagogical models they adopt should be guided not only by the benefits and limitations of mobile technologies, but also by the way in which these technologies operate. Integrate in the social and cultural fabric.

The use of wireless, mobile, portable and portable devices is gradually increasing and diversifying in all sectors of education and in developed and developing countries. It is moving from small-scale short-term trials to a wider and more sustainable deployment. Recent publications, projects and essays are used to explore the possible future and the nature of mobile education. This chapter concludes with an examination of the relationship between the challenges of a rigorous and appropriate evaluation of mobile education and the challenges of integrating and integrating mobile education into formal institutional education. Mobile technologies are also changing the nature of work (the driving force behind education and training), especially knowledge. Mobile technologies are changing the balance between training and performance support, especially for many knowledge workers. This means that "mobile" is not just a new adjective that qualifies the timeless concept of "learning"; rather, mobile learning is emerging as a completely new and distinct concept along with the mobile workforce and the connected society. Mobile devices not only create new forms of knowledge and new ways to access them, but they also create new forms of art and performance, and new ways to access them (such as music videos designed and sold for iPod). Mobile devices also create new forms of commerce and economic activity. Therefore, mobile learning is not a matter of "mobile" as we have already understood it, nor of "learning", as we have already understood it, but it is part of a new mobile conception of society. (This can be contrasted with technology-assisted learning or sustained technology, giving the impression that technology is doing something to learn).

2 Related Works

2.1 Learning Through Mobile

Currently, mobile learning uses both portable devices and mobile phones, as well as other devices that use the same set of functions. Mobile learning with handheld devices is obviously relatively immature in terms of technology and pedagogy, but it is growing rapidly [1].

In addition, the definition of mobile learning can highlight the unique attributes that place it in informal rather than formal learning. These attributes place a large amount of mobile learning at variance with formal learning (with its cohorts, courses, semesters, assessments and campuses) and with its monitoring and evaluation regimes. The difference also raises concerns about the nature of any large-scale and sustainable deployment and the extent to which the unique attributes of mobile learning can be lost or compromised. If we consider mobile learning in a broader context, we must recognize that mobile, personal and wireless devices are radically transforming social notions of discourse and knowledge and are responsible for new forms of art, employment, language, commerce, deprivation and crime, as well as learning. With increasing popular access to information and knowledge anywhere and anytime, the role of education is challenged, perhaps especially formal education, and the relationship between education, society and technology is more dynamic than ever. This chapter explores and articulates these issues and the links that link them specifically in the context of a broader and more sustainable development of mobile learning.

2.2 Game Based Learning

Young people have been playing computer and online games with enthusiasm and perseverance since the 1960s and 1990s, respectively. Now, computer and online games are more prolific and popular than ever. Through mobile learning devices and cloud computing, educational institutions are benefiting from the same determination, enthusiasm and perseverance as students when they play. Gareth Ritter, a Cardiff professor, explains how many children in this school play Call of Duty, if they fail on one level, they will not give up, they will continue to do so, we must take this to the classroom [2] Game-based learning seems to be the most effective way to teach students the basic concepts that would have already been learned through repetition and written exercises.

2.3   Virtual and Remote Learning Platform

The new and previously unimaginable possibilities of learning environments also result from the fusion of our physical and virtual worlds. The class is no longer limited to the one existing in a physical education institution; It can be anywhere the student chooses. This is the idea encapsulated by the VLE. At the forefront of this idea are the virtual and remote learning platforms. As mentioned above, VLEs are electronic education educational systems based on online models that mimic traditional in-person education. ELVs can include most learning environments, from virtual learning platforms such as MOOCs to virtual worlds such as those used for learning games [3]. Virtual and distance learning platforms can provide an interesting and interactive learning environment to any student, within or outside the traditional educational infrastructure.

2.4   M-Learning Frame Model

In the FRAME model, it is considered that mobile learning experiences exist in an information context. Collectively and individually, students consume and create information. The interaction with information is mediated by technology. It is through the complexities of this type of interaction that information becomes meaningful and useful. In this information context, the FRAME model is represented by a Venn diagram in which three aspects intersect (Fig. 1).

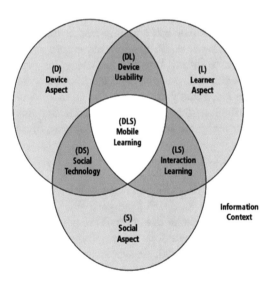

Fig. 1. The Frame Model

The three circles represent the apparatus (D), the apprentice (L) and the social spectrum (S). The intersections where two circles overlap contain attributes that belong to both aspects. The attributes of device usability (DL) and social technology (SD) describe the possibilities of mobile technology. The intersection called learning interaction (LS) contains teaching and learning theories that emphasize social constructivism. The three aspects overlap at the main intersection (DLS) in the center of the Venn diagram. Hypothetically, the main intersection, a convergence of the three aspects, defines an ideal mobile learning situation. By assessing the extent to which all FRAME model domains are used in a mobile learning situation, professionals can use the model to design more effective mobile learning experiences.

Towards More Effective Mobile Learning Environments

Even if students do not share the same physical environment, they can use mobile devices to share aspects of their personal and cultural lives. To solve problems specific to their situation, students can easily choose from a seemingly unlimited amount of data. The Internet marked the beginning of an era in which information became easy to access and easy to publish. Now, students must acquire the skills and tools to navigate this growing body of information [4]. Mobile learning allows students to interact with additional tools such as text messages, mobile Internet access and voice communications, all over wireless networks. Although this medium can be hampered by a low bandwidth and limited input and output capabilities, it has some clear advantages:

- Wireless and networked mobile devices can provide students with access to relevant information when and where it is needed. Mobile students can travel to unique locations, physically or virtually through their mobile devices.
- The ability to access a variety of materials from anywhere and at any time can provide multiple clues for understanding and retention.
- Learning in specific contexts can provide authentic cultural and environmental clues to understand the uses of information that can improve coding and retrieval.
- Well-implemented mobile education can help reduce the cognitive load for students. Although it is difficult to determine how to group information, presentation schemes and different amounts of information can potentially help students retain, retrieve and transfer information when necessary.

3 State of Mobile Learning

Nowadays, mobile technologies, initially marketed mainly as communication and entertainment devices, now play an important role in economies and society in general. Mobile devices have been used in almost every area, from banking to politics, and are currently used to increase productivity in many industries. As these devices become increasingly popular around the world, there is a great interest in mobile learning. Students and teachers are already using mobile technologies in a variety of contexts for a wide variety of teaching and learning, and key educators - from ministries of education to local school districts - are experimenting with support policies to promote learning innovative mobile, both formal and informal. A large number of experts interviewed for this report believe that mobile learning is now on the verge of a more systematic integration with education inside and outside schools. The decisions made today will fundamentally influence the character of mobile learning in the coming years. To help set the stage for these decisions, the following sections describe some of the most prevalent trends in mobile learning to date. These include innovations in formal and informal education, lifelong learning and educational technology [5].

Formal Education

The presence of mobile devices in formal education systems is growing. Worldwide, two of the most popular models for mobile learning in schools are one-to-one (1:1) programmers, where all students receive their own devices at no cost to students or their students; and the Bring Your Own Device (BYOD) initiatives, which are based on the prevalence of student-owned devices, with schools that provide or subsidize devices for students who cannot afford them. As expected, the 1:1 model tends to be more prevalent in poorer countries and regions, while the BYOD strategy is usually implemented in wealthier communities where mobile is almost ubiquitous.

Bring Your Own Device (BYOD)

A viable way to achieve a 1:1 environment is to have students use the mobile devices they already have. This model, known as BYOD, is already causing a major shift in higher education and distance education by allowing more students to access course materials through mobile technology. With increased access and mobile ownership, BYOD is promising for students around the world, although it may seem radically different in all regions and countries. Although the strategy has been more popular in countries and communities where the ownership of smartphones and tablets is widespread, students and educators have also found ways to take advantage of the less sophisticated technologies that students possess. The Nokia MoMath project in South Africa, for example, uses SMS (Short Message Service) functions in standard mobile phones to provide students with access to content and mathematical support. As BYOD changes the cost of school hardware to the student, it puts additional strain on bandwidth, a fundamental consideration for the infrastructure of mobile learning initiatives. Schools or governments that implement BYOD programs must also have a strategy to provide devices to students who cannot afford them, either by purchasing devices for students or by subsidizing their purchase. Other problems include security, privacy, adequate professional development for teachers and a digital divide between students with advanced devices and those with less powerful devices or none at all. For these reasons, examples of successful BYOD initiatives, particularly in elementary and secondary schools, are limited.

Informal Education

Mobile learning has developed, to a large extent, outside of formal educational contexts, and the vast majority of mobile learning projects are designed for learning to inform.

Informal Education

Continuous learning is defined as uninterrupted learning in different environments, including formal and informal settings. In the ideal scenario of continuous learning, a student opportunistically uses different types of technologies, taking advantage of the unique advantages of each one, the mobility of a smart phone, for example, or the upper keyboard in a desktop, to maintain continuity. Historically, there has been a significant division between formal learning in the classroom and informal learning that occurs in the home or in a community setting. Many experts are exploring how mobile learning can help overcome this barrier and close the gap between formal and informal learning.

Educational Technology
The latest innovations in mobile technologies have focused on the creation of digital content, mainly in the form of digital manuals accessible through electronic readers, and the development of mobile applications (applications) and software platforms to access resources through mobile devices [6].

Mobile Apps
Markets for mobile applications have provided a completely new distribution mechanism for the content, stimulating a substantial investment in the development of software for mobile devices. Educational applications are already experiencing significant growth in developed countries. These applications provide new tools for educational activities such as annotation, calculation, composition and content creation. A recent study found that in 2011 270 million educational applications were downloaded, an increase of more than ten times since 2009. Although a small number of educational applications are associated with the objectives of the curriculum and are designed for use in the classroom or at home, most of them are mainly for informal learning. However, as more and more students use mobile devices in formal education environments, applications will likely become an important part of the mobile learning ecosystem. Not only can developers now bypass institutions and sell content directly to students, but students, teachers and schools can make small additional investments in small amounts of content.

The advent of mobile technologies has created opportunities for the provision of learning through devices such as PDAs, mobile phones, laptops and tablets (laptops designed with a handwriting interface). Collectively, this type of service is called m-learning. While m-learning can be seen as a subset of e-learning (which is content delivery and web-based learning management), the emerging potential of mobile technologies tends to indicate that the learning framework m- Learning also has direct links to the flexible learning model "enough, just in time, just for me" (see Fig. 2) and, therefore, it is only one option among others it can be adapted to individual learning needs.

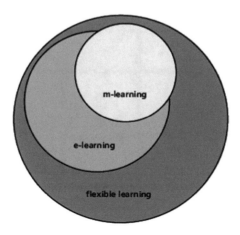

Fig. 2. M-learning as a subset of E-learning

4 The Future of Mobile Learning

With more than 5.9 billion mobile subscriptions worldwide, mobile devices have already transformed the way we live. But even though people around the world rely heavily on mobile technology, educators and policymakers have not yet taken full advantage of their potential to improve learning. The next decade and beyond could be transformative by integrating mobile technologies into formal and informal education to better meet the needs of students and teachers around the world. The following sections describe some of the technological advances that are likely to have a greater impact on mobile learning in the future, and highlight key areas in the development of mobile learning over the next fifteen years.

Mobile learning appears as one of the solutions to the challenges of education as seen in Fig. 3. With a variety of tools and resources still available, mobile learning offers more options to personalize learning. Mobile learning in the classroom often involves students working interdependently, in groups or individually to solve problems, work on projects, respond to individual needs and allow students to make their voices and choices heard. With access to both content at anytime and anywhere, there are many opportunities for formal and informal learning, both inside and outside the classroom. The study shows that laptops, mobile tablets, iPod touch and iPads are very popular. They are used to collect student responses (clickers), read e-books and websites, record reflections, document field trips, collect and analyze data, and much more. One of the reasons for accepting mobile learning is that it uses devices: The future of mobile learning depends to a large extent on the level of social acceptance it receives. On the other hand, users in developing countries have the same need to be mobile, accessible and affordable, as do developed countries. The true meaning of M-Learning is its ability to make learning mobile, away from the classroom or the workplace. These wireless and mobile technologies provide learning opportunities for students who do not have direct access to learning in these places. Many students in developing countries have difficulty accessing the Internet or have difficulty providing technology for learning in an e-learning environment. Mobile devices are a cheaper

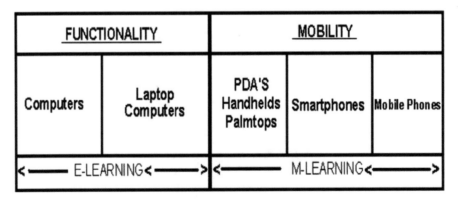

Fig. 3. Functionality and mobility definition of mobile learning

alternative to traditional E-Learning devices, such as PCs and laptops. The reason why mobile learning does not move to get out of the status of your project and take its place in conventional education and training is well known. Telecom operators do not consider mobile learning as a valuable and attractive revenue stream. Progress is being made in a wide range of mobile applications, but education and training are lagging behind [7].

5 Conclusion

The use of mobile technology in education is a recent initiative due to the availability and rapid progress of mobile devices such as smartphones, PDAs and handheld devices. Recently, there have been numerous research studies and applications of mobile learning in formal and informal learning. The chapters in this book present some of these studies and recent projects on mobile learning in education and training. This study has focused on the impact of technology on the future of education. The impact of technology on education, however, is not reserved for the future; Technology is innovating ideas and methods of education. It seems like a very exciting time to be a student, but the best technology discussed in this report is that it allows us all to be students, anywhere, at any age and at any time. There is no doubt that the world of higher education in the world will undergo a great transformation. With newer and more affordable technologies, classrooms around the world are changing. Through better understanding and use of these incredibly powerful new revelations in the field of educational technology, we can prepare future generations for what the future holds.

Acknowledgements. This Research was supported by the MSIP (Ministry of Science, ICT and Future Planning), Korea, under the ITRC (Information Technology Research Center) support program (IITP-2018-2013-0-0087) supervised by the IITP (Institute for Information & Communications Technology Promotion).

References

1. Gayeski, D.: Learning Unplugged: Using Mobile Technologies for Organizational and Performance Improvement. AMACON - American Management Association, New York (2002)
2. Attewell, J.: Mobile communications technologies for young adult learning and skills development (m-learning) IST-2000-25270. In: Proceedings of the European Workshop on Mobile and Contextual Learning, The University of Birmingham, England, The University of Birmingham, England (2002)
3. The Future of E-Ducation: The Impact of Technology and Analytics on the Education Industry, Gold Mercury International (2013)
4. Bruner, J.: The Process of Education: A Searching Discussion of School Education Opening New Paths to Learning and Teaching. Vintage Books, New York (1960)
5. Keegan, D.: Mobile Learning: The Next Generation of Learning. Distance Education International (2005)

6. Churchill, D.: Towards a useful classification of learning objects. Educ. Tech. Res. Dev. **55**(5), 479–497 (2007)
7. Mehdipour, Y., Zerehkafi, H.: Mobile learning for education: benefits and challenges. Int. J. Comput. Eng. Res. **3**(6), 93–101 (2013)

Design and Evaluation of Soil pH Iot Sensor Attribute for Rice Agriculture in Central Africa

Symphorien Karl Yoki Donzia[1], Hyun Yeo[2], and Haeng-Kon Kim[1(✉)]

[1] Daegu Catholic University, Gyeongsan, South Korea
{yoki10, hangkon}@cu.ac.kr
[2] Sun Cheon National University, Suncheon, Korea
yhyun@sunchon.ac.kr

Abstract. The Iot platform is a crucial middleware that allows us to walk safely in the field of IoT agriculture. The Rice Intensification System (SRI) uses less water and improves rice yield through the synergy of several agronomic management practices around the world. Rice is a staple crop for many countries and is one of the three main food crops in Central Africa. Environmental stresses such as drought, salinity, disease, acidity and iron toxicity can limit growth and performance. In Central Africa, rice is grown mainly in the lowlands. Africa consumes 11.5 million tons of rice per year, of which 33.6% is imported. Import trends showed only a slight decrease of 55 000 tones, while rice production increased by 2.3 million tones. And Central Africa's average self-sufficiency reached 37.9%. This is new way of more production in agriculture sector for saving a life have to be researched. The most natural one being the use of different machine in farm by connecting different sensors, connected devices, developing intelligent breeding systems as much as possible. In the paper, an architecture based on IoT agriculture is proposed after analyzing of soil organic. This work includes the expansion of rice management scales, including human resources, water and agricultural land; response strategies to irrigation during the drought; and the application of several agricultural water resources. Finally, it presents a architecture for IoT agriculture services based on cloud sensor infrastructure. Our results show that rice should be planted in lowland rice production systems in Central Africa.

Keywords: Farming · Information technology · IoT · Wi-fi
Agriculture and pH

1 Introduction

Rice is an important staple for much of the world's population and the largest consumer of water in the agricultural sector [1]. Global food security remains highly dependent on lowland irrigation rice, the main source of rice supply [2]. Fresh water for irrigation is increasingly scarce due to population growth, increased urban and industrial development and decreased availability resulting from pollution and depletion of

© Springer Nature Switzerland AG 2019
R. Lee (Ed.): ACIT 2018, SCI 788, pp. 37–47, 2019.
https://doi.org/10.1007/978-3-319-98370-7_4

resources. Asia contributes more than 90% of world rice production using more than 90% of total irrigation [3]. It is estimated that by 2025, 15 million of the 130 million hectares of irrigated rice in Asia could experience a "physical water shortage" and about 22 million hectares of dry season rice could suffer from "economic shortages". of water. [4] Rice is a very important and valuable crop for Taiwan's economy, he gives more than 1.73 million, and farmers should be able to improve their rice production while improving the quality of their soil and environment, reducing its limited fresh-water supply requirements. SRI's limited testing in Taiwan confirms that high grain yield can be achieved in the first growing season. Within this period, the productivity of labor, capital and Irrigation water productivity is high, whereas crops grown during the second season are exposed to a higher incidence of damage by pests, diseases and weeds, and floods create a greater risk of yield losses and reduce the potential operating margin for labor, capital and productivity gains of irrigation water [5]. In this context, this study was carried out using two thirds of the SRI practices mentioned above during the dry season, since knowledge about adaptation, growth and water saving is still limited to date. Understanding the effects of different irrigation regimes on root growth and the physiology of rice plants is essential to increase the productivity of water and rice crops, especially when using some SRI attributes for rice cultivation. Therefore, the objective of this research was to evaluate water productivity, crop growth and yield components of the two rice varieties using SRI management practices under different irrigation regimes. We real example of Rice is an aquatic plant, which is grown mainly in heavy and neutral soils and is also grown in lightly textured soils. And it is grown in loamy sands in Punjab or in Andhra Pradesh. The alluvial, red, lateritic or lateritic and black soils, which the rice prefers a slightly acidic pH, but which can grow in a pH range of 5 to 8 due to its better adaptation, are grown in extreme conditions. Soil conditions such as acid soils of Kerala peat With the exception of coastal saline soils, occurs in heavy rains, saline or alkaline soils negatively affect the growth and pro-ductivity of Rice.

The processes that decrease soil pH in agricultural systems are removal of base cations from the soil system (generally through plant uptake and removal of vegetation, or leaching), and application of ammonium-based nitrogen fertilizer at rates in excess of plant requirements. Liming cause soil pH to increase, though its effectiveness varies with soil texture, type of lime, and amount applied. This combination of variable loss and gain complicate detection of long term trends in soil pH. Here we describe the design of a soil monitoring network to characterize the current status and future trends in soil pH for the Central Africa (Fig. 1).

Current Issue; Ounce agricultural land resources are converted to non-agricultural uses, their productivity can hardly be recovered even when they are returned for agricultural purposes. African as a continent, and poisoning the world in general, the potential is phenomenal. Unfortunately, however, this potential remains largely untapped. Training is a noble and vital profession because without agriculture, the world is starving. Unfortunately, despite the great role farmers play in our lives, most farmers around the world continue to face great challenges in their daily lives. Most farmers in Africa today are small farmers or subsistence farmers who grow and raise animals solely to feed themselves and their families. In addition, most live in rural and

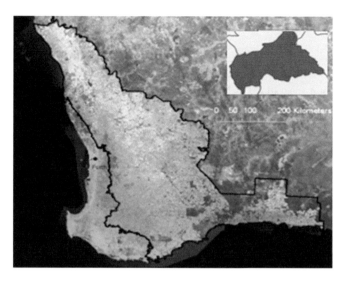

Fig. 1. Generally defined as the land cleared for dryland cropping systems that receive less than 500 mm annual rainfall

suburban communities. In general, the main problem is that lack of information remains the number one problem that most smallholder farmers in Africa are facing today. And poor financial support also means the lack of financial support systems to enable farmers to grow, expand and maintain their returns. What follows is the lack of access to fermenters; this falls under the wrong financial system mentioned above because farm land has become so expensive in Africa. And the transport of the poor; this is a major problem not only for agriculture, but also for the economy in general throughout Africa. Most agricultural products in Africa are simply wasted in remote areas and this is mainly because farmers find it very difficult to sell their agricultural products to the market. And last on the poor markets; the farmers market has become one of the most serious problems facing Africa, which is currently affecting the lives and living standards of millions of people.

Our Goal; Our goal is to find new direction and innovative way to feed a fast growing population. This is to allow the first to see the view of the farm seeing a visualization not only bad in terms of images, but also in terms of the state of the farm, such as the pH of the soil, the soil moisture varying on closes for different periods. Time to get that view everywhere, not always or only when they are on the farm, but even when they are gone.

Our Motivation; The relative humidity and soil, the pH level and the soil temperatures that will be extremely useful when it comes to spreading wine next season or whether it is or not, it's worth adding Irresponsibility to some crops if you do not observe differences in crop yield due to 20% and moisture. It may not be worthwhile to water these particular crops, saving time, money and ultimately making the farm more productive. So we believe the technology will benefit small farmers around the world.

2   Problem Soils and Management System of Rice

The problematic soil is one where which can't be cultivated, but it must adopt certain specific and economic management practices to alleviate soil restrictions to a certain extent and at the same time choose the most suitable crops for the situation. Rice is the typical crop that [6] requires a different soil environment to grow. Since rice is submerged, soil restrictions are partially relieved. Most rice cultivated in a problematic nature and their cultivated nature which is suitable in these soils. As a result, the main problematic soils under which rice is grown are informed by their management. These problematic rice soils include; Sodic soils, saline soils, acid soils and peat bogs. Program analysis is the process of analyzing the behavior of a computer program. Program analysis has a very widely application range, it provides support for compiler optimization, testing, debugging, verification and many other activities [10].

2.1   pH (Soil Acidity) in Reaction

To indicate the degree of acidic or alkaline soil: presented as pH (calculated: $-\log$ [H +]). The validity of the different nutrients in the soil also, the solubility of the toxic substances, influence the physiological reactions of the roots of the plants and the microbe.

2.2   pH Calibration: pH = Lower

In order to obtain the required amount of lime in the pH adjustment of the soil you should look for the requirement of lime (LR). The higher the clay content of the buffer capacity of the soil and the organic matter content is a higher requirement for the lime is increased. Needs of the first lime: Obtain the required amount of calcium carbonate to adjust the target pH (pH) of 6.5. Siyongryang of slaked lime, Goto fertilizer. If the Siyongryang limestone fertilizer is more than 300 kg/10a, it fails in the harvest, so we have some concerns. And the match over 2–3 years Zoom Excessive draft: Deficiency of iron, manganese, copper, zinc caused, decreased phosphate usage Inhibits boron absorption [7] (Table 1).

Table 1. Effects of siliceous fertilizer flood

Pass over non-siliceous fertilizers	Non-siliceous fertilizers		
	Towards the wetlands	Alpine mountains	Not cold
100	151	120	110

pH correction: increase in pH

Standard Fertilization: a base of 150 kg/10a. And Black Fertilizer is almost Soil Silicate Active Silicic Acid ($SiO2$) 157 mg test/fit kg. Which gives Silicate fertilizer (kg/10a) = (157 - Plot of $SiO2$) $\times$ 4.2? application. The most important thing is that the silicate is not specified separately. In 3 years the trial period, before working the fumigation, penetrating fertilization.

2.3   Characteristics of Soil Organic Matter (Corrosion)

It is presented in both tables; first, it is the physical improvement that binds to form the soil in reduced bulk density. (The air-oxygen offers a smooth and good root sweep). The second is the improvement of the chemical resistance with which the nutrients increase in the capacity of adsorption of the soil (Fig. 2).

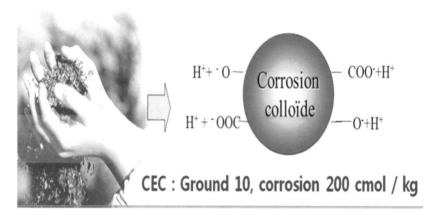

Fig. 2. Ground(-) Increase the amount of fillers that increase the nutrient retention capacity and become moderate nutrients to avoid excessive interference.

Management of soil chemical sciences (a) Fertilization by prescription of Soil Testing - The amount of food available to feed the soil crops identified. pH, organic matter, available phosphorus, interchangeable cations, active silicic acid, need for lime, etc. - Fertilizer is recommended to examine the nutrient content of the soil (b) Scientific fertilization by a field of diagnosis – [8] The state of nutrients in the soil according to soil management practices - The dose provides the necessary nutrients through timely analysis at the site (Table 2).

Table 2. Soil chemical changes

Year	pH (acid)	Organics (G/kg)	Available phosphorus (Mg/kg)	Exchangeable cations (cmol/kg)			Effective Silicate (Mg/kg)
				K	Ca	Mg	
'99	5.7	22	1.36	0.32	4.0	1.4	86
'03	5.8	23	141	0.30	4.6	1.3	118
'07	5.8	24	132	0.29	4.7	1.3	126
'11	5.9	26	131	0.30	5.1	1.3	146
'15	**5.9**	**26**	**140**	**0.30**	**5.5**	**1.2**	181
Appropriate range	5.5~65	20~30	80~120	02~03	5.0~6.0	1.5~2.0	157 more

2.4   Fertilizer of Nitrophenes for a Large Cantidad de Mala Calidad

Ubrique was delayed and extended from the cap period. Unknown rice growth: A lot of rice to distribute to photo-synthesis, and times and reduces photosynthetic efficiency by water (Fig. 3).

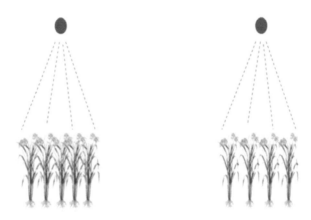

Fig. 3. Ambiguity rice growth

➡ Reduction of the lip of the fabric ➡ Degradation

Saline soils are those soils contains sufficient soluble salts to interfere with crop growth. soils remain flocculated and the hydraulic conductivity is more or equal to non saline soils. When Soil pH is <8.5, electrical conductivity is > 4.0 dSm-1 at 25 °C and Exchangeable sodium percentage is <15 Cmol(P+) kg–1. These soils have little physiological water availability due to the high osmotic potential, low aeration and toxic effect of sodium carbonate, sodium chloride and sodium sulfate. Although rice crops are considered salinity tolerant, no variety of rice can withstand the adverse effect of high salinity throughout its life cycle. Due to submersion in most rice soils, it gives tolerance to salinity. Maintain flooding throughout the life cycle of rice, which is suggested to improve rice harvest under saline soil conditions. Selection of salt tolerant rice varieties based on salinity levels and water depths. Deep ploughing be-fore submergence, heavy pre-sowing irrigation, high seed rate, closer spacing, green manuring with Dhaincha, use of organic amendments, placement of N fertilizer below soil surface, split application of N [9] (Table 3).

Table 3. How to improve soil types

Floor type	How to make a large amount
Habitually	State of mind, organic and silicate test
Rice	Clay, organic matter using zinc silicate, potassium phosphate, stocks
American rice	The feelings try to walk the organic phosphate
Wet rice fields	**Clay, drainage, try silicate, potassium stocks**
Non-chloride	Customer sat, sea water, using the organic plaster of zinc

The major soil orders for rice cultivation are alfisols, entisols, inceptisols and ultisols. The most important feature [8] of rice land is submersion for at least part of the growing season. Immersion leads to changes in soil pH, which tends to reach neutrality in acidic and sodic/calcareous soils. Submergence increases the availability of Fe, Mn and P but leads to loss of N by denitrification and reduced nitrogen use efficiency. Lime application play greater role to reclaim acid soils. Ground rock phosphate is better alternative to water soluble P fertilizers in acid sulphate soils. Application of this system is important for sodic soils and selection of tolerant varieties with organic amendments are important for better growth of rice under saline soils. And Major problem soil under which rice is grown are sodic soils, saline soils, saline-sodic soils, acid, red and lateritic soils, acid sulphate soils and peat soils. So we need to combined with IoT Farm, in the second part of our work, Most generic technology components to address these issues are expected to be available. As Our objectives were to evaluate soil profile chemical characteristics more technology so by 2030, Africa wild need to feed 1 billion people during this work, We are going to demonstrate how low-cost technology can make small-scale agriculture more productive to help meet the challenge of feeding a increase Central Africa.

3 Smart Farming Agriculture

Agricultural transformation integrates two main processes. The first is to transform or modernize agriculture by increasing the productivity and functioning of farms as modern enterprises. The second is to strengthen linkages between farms and other economic sectors in a mutually beneficial process, where agricultural production supports manufacturing (through agribusiness) and other sectors support agriculture by providing inputs. [10] Earlier work has also focused on precision farming applications for irrigation, variable seeding, nutrient application and others. There has been previous work on the development of technology to enable precision farming. Researchers have built specialized sensors to measure nutrients [10], water levels and other similar sensors, and build on this work. Complementary to this work package, as it facilitates the automation of data collection using these sensors and enables precision culture systems. Agriculture is one of the oldest industries and technologically it is probably one of the most backward. Technical Community and IT community are looking for ways to solve this problem of agricultural technology. So, the different ways that people are trying to help a farmer in one of the ways that we look for in our previous work is that we need a way to double lead in agriculture by 2030; we need an interruption, how can we get there?

We address the following objectives like availability that the platform has a negligible downtime. The Capacity need admit sensors with very variable requirements; PH sensor that informs a few bytes of data to drones that send gigabytes of video. And also cloud connectivity as several agricultural applications, such as crop cycle prediction, planting suggestions, agricultural practice tips and more. And the last one is the data update that the data from a farm condition sensor can do to the farmer.

3.1   Overall Architecture of System Based on IoT Agriculture

The architecture model proposed as shown in Fig. 4. It consists of the five modules: (1) Sensor module. (2) Mobile application module. (3) Cloud module (4) Analysis and knowledge (5) User interface module. The Sensor Kit is a portable IoT device with floor and environmental sensors. [11] The mobile application module provides an interface for users.

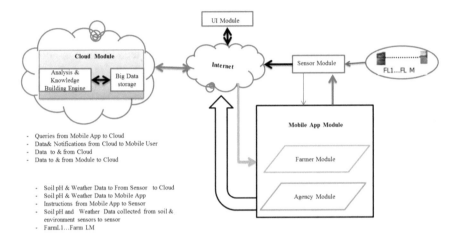

Fig. 4. Overall architecture of the system

A. Internet of Things

The current use and potential of IoT technologies to solve development problems, highlighting a series of specific cases where IoT interventions help solve some of the most pressing problems in the world. IoT can reach billions of people living in a developed world and accelerate growth and development of income accordingly. IoT can be better understood as a set of technologies that can be used together to achieve interesting objectives, and can be defined in terms of contributing technologies, including the use of sensors. Any IoT system includes a set of functions and types of connectivity. The Internet is the global system of interconnected computer. The IoT is expected to be a powerful driver that will transform agriculture and food into smart webs of connected objects that are context-sensitive and can be identified, sensed and controlled remotely [8]. This is contemplate could formal agro-food active in groundbreaking ways. Arise from new control device. We consider honestly that IoT will be a real time changer in agriculture or general food chain that increase productivity. Like in example, IoT will help farmers to make towards data-driven farming carry with resolution making tools from time or accurate running data. As a result, farms could withdraw from traditional supply-oriented, unidentified approach to cost-based, supply continuously are equal. Farms or food supply chains could shows the advantage of self-adaptive systems in which smart are cover the farm equipment or slight intervention by humans.

3.2   Properties of IoT Applications in Agriculture

To solve the problem of food security and water security through sustainable agriculture, the solution should provide additional information/services, such as third-party agricultural services, microfinance services, [12]. for farmers. It should also provide a central repository for a variety of information, such as traditional sustainable agricultural techniques, crop diseases, etc. from various source, allow interactive agriculture, facilitated access to users in various devices, such as telephones. IVR, computers and kiosks, in addition to providing multilingual support of traditional practices with modern value. The system must meet the following requirements:

1. Robust models: The characteristics of the agricultural sector, such as diversity, spatial and temporal variability in complexity and uncertainties, must be taken into account when developing appropriate types of products and services.
2. Scalability: The size of the farms varies from small to large and, therefore, the solutions must be scalable. Testing and deployment are done in stages and, therefore, the architecture must be able to expand incrementally with fewer resources.
3. Affordability: Affordability is the key to success. The cost must be appropriate with substantial benefits. Standardized platforms, tools, products and services can reduce costs by increasing volume [13].
4. Sustainability: The point of sustainability is active just because of acute economic harassment or aggressive global competitiveness.

4   Evaluation and Assessment

4.1   Evaluation

As for the sensors, they are usually placed on a farm destined to bring new technologies marketed as safety devices. In addition to IoT services, IoT applications deserve special attention. The following comparisons were made from the following research works discussed in the related work section of this manuscript.

- Automatic increased Soil pH based on IoT farming
- Development of Smart Safety Crop Sensor
- Overview of Current farming agriculture based on IoT technology
- Development of smart agriculture with IoT Gateway of area coverage algorithms.

Figure 5 [14] shows a service which used to develop applications, while applications are used directly by farmers. You can see that services are used to develop applications, while applications are used directly by farmers. The maximum speed was set to 10 m/s and the altitude was set to 20 m. Figure down plots the time taken to complete a flight with the two algorithms in different area.

Apart from sensors on hardware and software side, sensors are of two types: Built-in sensors and External sensors. Table 4 shows the some sensors and its potential application and advantages in agriculture, which helps in generating and providing agricultural data for further management and analytics etc. Big data analytics wide awake farmers from experiential problems like dryness environment etc. [14].

Fig. 5. Smart Agriculture based on IoT Gateway

Table 4. Farmers agriculture decision making based on sensors data

Decision making	Based on data
When and how to irrigate a field	Soil moisture data weather predictions data crop health data
Planting and harvesting decisions	Yield data weather data
Fertilizer applications and prescription	Soil nutrient density data plant/pest diseases data

4.2 Assessment

To control the environment in an intelligent agriculture, different sensors are used that measure the environmental parameters according to the needs of the plant. We can create a server in the cloud to access the system remotely when connecting through the IoT. Smart Farming uses technologies, sensors and data from the Internet of Things (IoT) to maximize crop yields and reduce waste. This report examines the roles of the many suppliers in the value chain, the suppliers of agricultural equipment. It also provides detailed perspectives on the future of [15] smart agriculture for strategic and technological planning relevant to mobile network operators, other connectivity providers, sensor manufacturers and software developers specializing in agricultural solutions.

5 Conclusion

The further occupations will focused on based framework U- Agriculture that involve IoT to sprint parallel in high increase world. The IoT agriculture implementation are building it feasible for ranchers to collect significant data. Huge owners and little farmers need to kown the potentiality of IoT market for agriculture by installing smart advanced technologies for the expansion competitiveness. The study showed that not all specific attributes of SRI administration are required to have a positive effect on plant growth, increased yields and increased water productivity. The challenges to sustain or maintain rice production are drastically increasing as fresh water for agriculture is sought of by other sectors. SRI offers the opportunity to reduce world hunger and sustain ably manage world water resources; however, it merits a thorough comprehensive research program to unlock its full potential.

Acknowledgement. This research was supported by the MSIP (Ministry of Science, ICT and Future Planning), Korea, under the ITRC (Information Technology Research Center) support program (IITP-2018-2013-1-00877) supervised by the IITP (Institute for Information & communications Technology Promotion).

References

1. Thakur, A.K., Mohanty, R.K., Patil, D.U., Kumar, A.: Impact of water management on yield and water productivity with system of rice intensification (SRI) and conventional transplanting system in rice. Paddy Water Environ, **12**, 413–424 (2014)
2. Yang, J., Zhang, J.: Crop management techniques to enhance harvest index in rice. J. Exp. Bot. **61**, 3177–3189 (2010)
3. Khepar, S., Yadav, A., Sondhi, S., Siag, M.: Water balance model for paddy fields under intermittent irrigation practices. Irrig. Sci. **19**, 199–208 (2000)
4. Tuong, T.P., Bouman, B.A.M.: Rice Production in Water Scarce Environment. Manilla, Philippines, International Rice Research Institute (2003)
5. Chang, Y.C., Chen, T.C., Hsieh, J.C.: Feasibility of system of rice intensification in Taiwan. Taiwan Water Conserv. **61**, 1–11 (2013)
6. Maier, N.A., McLaughlin, M.J., Heap, M., Butt, M., Smart, M.K., Williams, C.M.: Effect of current-season application of calcitic lime on soil pH, yield and cadmium concentration in potato (Solanum tuberosum L.) tubers. Nutr. Cycl. Agroecosyst. **47**(1), 29–40 (1996)
7. Schou, J., et al.: Design and ground calibration of the Helioseismic and Magnetic Imager (HMI) instrument on the Solar Dynamics Observatory (SDO). Sol. Dyn. Observatory **275**, 229–259 (2011)
8. Clark, M.S., Horwath, W.R., Shennan, C., Scow, K.M.: Changes in soil chemical properties resulting from organic and low-input farming practices. Agron. J. **90**(5), 662–671 (1998)
9. Saxton, K.E., Rawls, W., Romberger, J.S., Papendick, R.I.: Estimating generalized soil-water characteristics from texture 1. Soil Sci. Soc. Am. J. **50**(4), 1031–1036 (1986)
10. https://www.google.co.kr/search?dcr=0&biw=1920&bih=974&tbm=isch&sa=1&ei=iAnSW sGIIIqa0gTd7aygBg&q=Future+of+smart+agriculture+with+modular+IoT+Gateway&oq
11. Huang, D., Wu, H.: Mobile Cloud Computing: Foundations and Service Models. Morgan Kaufmann, San Francisco (2017)
12. Sumelius, J., Bäckman, S., Kahiluoto, H., et al.: Sustainable rural development with emphasis on agriculture and food security within the climate change setting: SARD-climate final report (2009)
13. Edenhofer, O., et al.: On the economics of renewable energy sources. Energy Econ. **40**, S12–S23 (2013)
14. Rehman, A., Jingdong, L., Khatoon, R., Hussain, I.: Modern Agricultural Technology Adoption its Importance, Role and Usage for the Improvement of Agriculture (2016)
15. Evaluation. http://www.beechamresearch.com/downloads.aspx

A Study on the Architecture of Mixed Reality Application for Architectural Design Collaboration

Kiljae Ahn[1], Dae-Sik Ko[1(✉)], and Sang-Hoon Gim[2]

[1] Mokwon University, Daejeon, Republic of Korea
{giljaean,kds}@mokwon.ac.kr, manfromwest@gmail.com
[2] Dongwoo E&C co., Seoul, Republic of Korea
gimsang@dwoo.co.kr

Abstract. The purpose of this study is to elucidate the utilization of mixed reality technology in the field of Architectural design. For this purpose, this study introduces the mixed reality development in architectural design and construction business and suggests design collaboration service platform composed of mixed reality service server, which is a web service system for authoring and managing mixed reality contents, and terminal application that implements it in mixed reality environment.

The result of this research is that mixed reality contents can be used in a mixed reality environment after storing and setting the mixed reality contents on a server, and a collaboration environment can be implemented using an MR device. In addition, proposing a simple mixed reality teleconference system using a relay system to support remote conferencing in mixed reality environment.

1 Introduction

The mixed reality(MR) technology is not a new technology which has been developed since 1966, and it has often appeared in many SF films in the past. However, it has not been popularized due to issues in visualization, ease of use, and price.

Recently, large influential companies such as Google, Microsoft, Samsung and Lenovo have joined the mixed reality market and are emerging as core ICT industries. Mixed reality wearable devices such as the HoloLens [1] of Microsoft and AR service development environment based on reality space such as Apple's AR kit [2] which developed from the overlap of existing simple information, can be an example.

Accordingly, a variety of applications utilizing these technologies are being produced, and companies such as IKEA can freely arrange three-dimensional objects such as furniture and interior space components in a real environment by simply placing their furniture at home. It is possible for the users to simulate the appearance (Fig. 1).

© Springer Nature Switzerland AG 2019
R. Lee (Ed.): ACIT 2018, SCI 788, pp. 48–61, 2019.
https://doi.org/10.1007/978-3-319-98370-7_5

Fig. 1. Ikea place application using ARkit (IKEA, 2017)

Mixed reality technology in space design has the possibility to help intuitive understanding among the various members who share work in the design stage by overlaying the existing results such as 2D drawing or 3D model data. Currently, such services are concentrated in applications that support 3D models developed through in-house application software and its software. Mixed reality applications that support universal 3D models include contents created by designers, are still at the stage of expressing a simple 3D model. The concept of this study is to improve the architectural design process by sharing the 3D models of construction companies, design offices, and customers in the mixed reality environment, centering on the MR Service Server providing the web service (Fig. 2).

In order to realize this concept, this study suggests uploading the architectural 3D data of users to the mixed reality service server, so that contents can be easily produced only by the setting value according to the purpose of viewing and to utilize the mixed reality environment by using the general mixed reality devices.

The architectural design collaboration environment proposed in this study aims to implement the following functions:

- Register a generic 3D model and manage it web service.
- PC applications that can deploy and manipulate 3D models in a virtual environment.
- Mixed reality mobile applications that can deploy and manipulate 3D models in a real environment.
- Communication supports the data exchange between mixed reality headset applications.

The web service converts the digital content of the users into contents usable in mixed reality headset applications. It also supports the registration to the account and the management of the registered digital models. PC applications deploy single or multiple 3D models in a virtual environment on the monitor and communicate with other users. In mixed reality headset applications, 3D models in desktop applications are placed in the real world and communication with other users are performed.

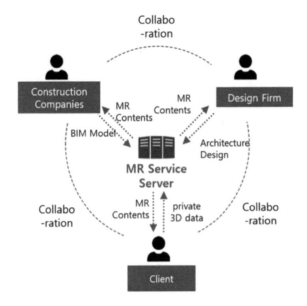

Fig. 2. The architectural design collaboration environment proposed in this study.

2   Mixed Reality Wearable Device and their Design Collaboration Applications

2.1   Mixed Reality Wearable Device

Mixed reality refers to the creation of information, such as an environment or visualization, by combining the virtual world and the real world. It uses the concept of interaction in real time between real and virtual existences.

In Comparing Virtual Reality and Augmented Reality, Virtual Reality places the user entirely in a digital location. It occludes the users' natural surroundings entirely. Whereas in Augmented Reality, such as AR glasses and mobile AR apps, the visible natural world is overlaid with a layer of digital contents. In Mixed Reality, on the other hand, virtual objects are integrated and responsive to the natural world. For instance, on the condition that a virtual building

Virtual Reality   **Augmented Reality**   **Mixed Reality**

Fig. 3. Introduction of mixed reality images from (Wired, 5, 2016)

is under a desk, as long as the user's gaze does not look away from the desk, it remains invisible (Fig. 3). HoloLens and Meta2 can be examples of Mixed Reality Headset.

As of March 2018, Mixed Reality wearable devices are being sold in small quantities to companies and developers. Typically, there is HoloLens from Microsoft and Meta2 from Meta co. as shown in Table 1 [3].

Table 1. Mixed reality headsets

	Hololens(Microsoft)	Meta2 (Meta company)
Standalone Unit	yes	no
Screen Resolution	1268x720	2560x1440
Viewing Area	30-degree	90-degree
Operation system	Windows 10(holographic)	Windows (separated)
Cost	3000 USD	1495 USD

Both system uses the positional tracking algorithm that fuses image features from the environment space, captured with onboard camera sensors, with the wearer's acceleration and angular movements, captured with an onboard inertial measurement unit (IMU). As a difference, HoloLens acquires spatial coordinates using the SLAM algorithm, and recognizes limited gestures relative to meta2. The service proposed in this study uses Microsoft's holographic lens, because it works solely without external devices and uses SLAM for spatial recognition.

2.2   Mixed Reality Design Collaboration Applications

Architectural design is currently being designed using BIM (Building Information Modeling). Through these, the work progress using the 3D model is being utilized as general business contents. To utilize these 3D contents, researches for introducing mixed reality technology centering on architecture companies are actively underway. Sample images of MR application in architecture design related businesses are shown in Table 2.

The application of mixed reality technology in the field of architecture is being developed for the architectural design and construction corporation mainly by the software developers of design related software, and the practicality test is underway.

Table 2. Sample images of MR application in architecture design related business

Architectural Design	Construction Planning	Field

Trimble co., Autodesk co.	Oyanagi Construction co.ltd	BAM

Autodesk and Trimble have developed a viewer for their design software. Through this software, the user can place the designed 3D model in the real space and review it as a 3D reduction model and review it at the real scale [4].

In cooperation with Microsoft, Oyanagi Construction Co. Ltd. is conducting development and field application experiments for the application of mixed reality to the field of construction simulation and field analysis of construction site [5].

The BAM is developing and applying experiments for the application of mixed reality technology to the management of buildings using BIM [6].

Cases are limited to specialized software companies and construction companies, such as software development for their own operations, and software that can only be utilized for specific design software.

3   Architecture of Mixed Reality Design Collaboration Service Platform

The mixed reality collaboration service platform is consisted of three parts; the online web server application, PC application and HoloLens headset application.

3.1   Web Service Application

The web service converts the digital content of the users into contents usable in mixed reality headset applications. It also supports registration to the account and the management of the registered digital models, as shown in Fig. 4.

In Fig. 5, the project management is the page of the top layer after log-in, and it creates an existing User MR project or a new project. In the Setting up project page, the service user performs settings needed for creating a Mixed Reality project. The template is selected according to the use purpose of the 3D model, and the pre-loaded assets to be used in the project among the MR Assets such as furniture and building materials provided by the service are also selected accordingly. The Setting up collaboration room field is located on the setting up project page, and allows other users who share the project to perform collaboration through the created project. Existing projects can also be changed

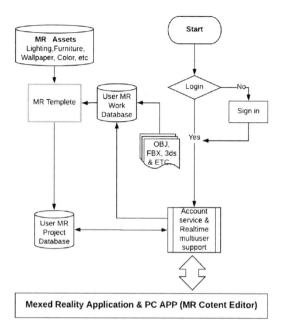

Fig. 4. Web service screen shots (above) and Web service server application architecture.

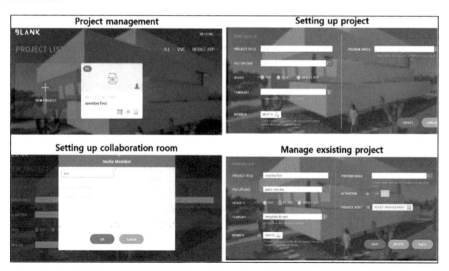

Fig. 5. Web service screen shots

by other users, and modified or added to existing MR assets through the same settings.

3.2   PC Application

PC applications complement the functionality of applications for Mixed Reality devices and implement mixed reality as virtual reality on the screen. This works similarly to the first-person game.

The main reason that we selected PC applications in implementing this service application, is because the modification of 3D objects in a mixed reality environment is difficult in a limited input environment of a mixed reality headsets. Therefore, functions such as the arrangement and modification of mixed reality contents are performed by the conventional keyboard and mouse through the PC.

As a result of evaluating the use of mixed reality application by eight architectural practitioners, it was found that all users were more comfortable to use than mixed reality devices in the placement and editing of objects.

In Fig. 7, the PC application chooses at login whether to manage the project as the Host or to participate in the registered MR space as a Guest. When the Host is chosen, the Main menu is the screen that appears after login through the PC application. The user has the authorization to the management of the MR project registered in the user's account, the management of the MR project registered in the local computer, and the management of the meeting attendee. When the user logs in as a Guest, only the project that the host user registers in advance for participation is shown to the user.

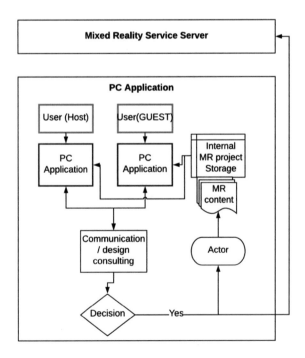

Fig. 6. PC application architecture.

The Registered Models corresponds to the MR Asset shown in Fig. 4, and it is possible to arrange or modify all of the users on the session.

MR projects can added contents or modified by users in the application is stored as an Actor, which is stored on the internal MR project and the service server (Fig. 6).

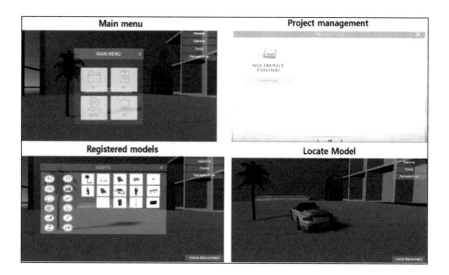

Fig. 7. Screen-shots from PC application

In this case, communication using the PC application is performed only between PC users. Screen-shots from the PC application is shown in Fig. 7.

3.3   Mixed Reality-Headset Application

Mixed reality headset application supports real-time placement and rotation of mixed reality contents and collaboration between multiple HoloLenses in the same network. In this case, the communication using the HoloLens is performed only between the HoloLenses. This is because the HoloLens uses its own sharing function to share spatial coordinates.

The HoloLens application chooses whether to manage the project as a Host when logging in or to participate in the registered MR space as a Guest. As the Host, the Main menu is the screen that appears after log-in through the HoloLens application. In the main menu, the Host can choose the project setup through the PC application, and can also communicate with Guests. The Registered Model corresponds to the MR Asset shown in Fig. 4, and it is possible to arrange or modify all of the users on the project. Changes made in mixed reality applications are stored in the web service server, so that changes in the project can be confirmed in the PC application.

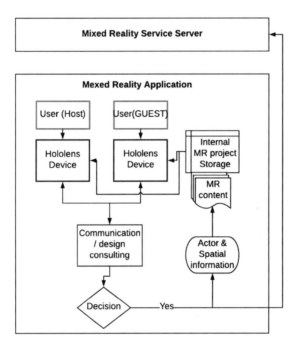

Fig. 8. Mixed reality application architecture

Mixed reality, unlike virtual reality, interacts with the real space, and objects are placed on actual surfaces such as desks and tables. Thus, HoloLens can use the spatial anchor function to allow multiple users to see the same model co-located in real space. The screen-shots from, mixed reality application (Fig. 9) and its architecture (Fig. 8) can be seen in Figures.

Fig. 9. Picture of using Mixed Reality application

4   Simple Remote Mixed Reality Collaboration System

The disadvantage of collaborative systems using mixed reality headset is that only the person directly using the wearing hardware can see the virtual content, thus users are unable to participate in remote collaboration or authoring. Mixed reality headsets, which are still in the developer product stage, are difficult to prepare in large numbers due to their high prices. In this chapter, a system is introduced that combines the system introduced in Sect. 3 with a hybrid reality headset + Digital Camera + image synthesizer, and a Spectator view SDK from Microsoft A.1.

4.1   Spectator View

Unlike meta2, HoloLens is able to send a live captured video from the headset to a Windows OS computer only via WIFI. Accordingly, Viewers who use computer screens without using a HoloLens, experiences a delay of several seconds. Microsoft Spectator view system allows others to see on a 2D screen what the HoloLens users see in their world without delay. It consists of a Spectator view rig and a computer with the video capture card (Fig. 10).

Using the spectator view SDK provided by Microsoft, we have made a collaborative space between users (up to four) of the holographic lens on the same local WIFI network. The spectator view hardware setup is shown in Fig. 11.

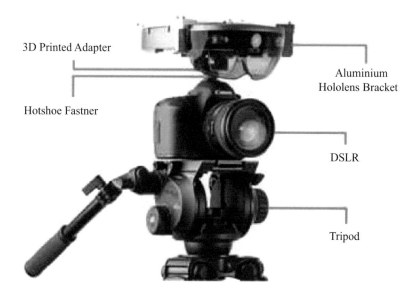

3D Printed Adapter

Aluminium
Hololens Bracket

Hotshoe Fastner

DSLR

Tripod

Fig. 10. The open-source version of spectator view hardware setup. Canon EOS 5D mark III, Blackmagic Design production camera 4K, Canon EF 14 mm f/2.8L II USM Ultra-Wide Angle Fixed Lens. https://docs.microsoft.com/en-us/windows/mixed-reality/spectator-view

Fig. 11. Spectator view hardware setup for this study. Canon EOS 5D mark IV, Blackmagic Design production camera 4K, ZEISS Milvus 2.8/15, and the desktop PC running the shared experience app and compositing the holograms into a spectator view video.

4.2   Real-Time Remote Relay for Remote Conference

In addition to the existing spectator view configuration, we implemented online remote collaboration function by adding video mixing PC and connect it with web-based remote relay services such as YouTube and Facebook. The overall Mixed Reality video conference system using spectator view is shown in Figs. 12 and 13.

Spectator view HoloLens transmits only spatial information such as the spatial model and spatial anchor to the Desktop PC, and synthesizes the virtual model to the HD video and sends it to the mixing PC. The mixing PC performs remote conference functions in conjunction with online real-time relay service,

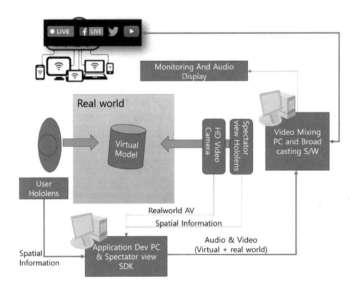

Fig. 12. Mixed Reality video conference system using spectator view

and at the same time, the conference video and sound is sent to the monitoring and audio display.

Fig. 13. Multi-party mixed reality collaboration test using developed system

When using video conferencing software (Skype), we confirmed that it is possible to configure a free conversation environment with no delay at a limited number of people. In addition, it was possible to relay to an unlimited number of people by using relay service such as YouTube, but it was possible to limit the conference with delays, one-way voice and text-based feedback. but in this case, a delay of few seconds occurs depending on the communication condition. However, internet connection speed and setup own broadcasting system the delay is expected to be reducible.

5   Conclusion and Future Work

Mixed reality collaboration systems are expected to grow strongly over the years. However, it is getting slower due to high prices and limited target devices and users. The prevalence of low-cost and user-friendly mixed reality collaboration services is an important requirement.

In this paper, we propose a system composed of Web service application, PC application and mixed reality headset application as a collaborative service system for architectural design support, and developed the architecture design and prototype software of each component. We also developed a prototype system for mixed reality teleconferencing for remote collaboration and proposed a low cost mixed reality teleconference system and its application.

However, the method proposed in this study is merely a relay of the screen displayed on a separate MR device, so that a conference attendant at a remote site cannot directly carry out operations such as movement and change, and relies on manipulation of a host user.

In the future, we will keep working on this prototype service application and try to combine VR, AR, and other mixed reality platform support to enable a multi-platform virtual conference environment.

Acknowledgements. This research was supported by The Korea Institute for Advancement of Technology – Grant funded by Korean Government (Ministry of SMEs and Startups, R0006121).

Appendix

A.1. Open-source software used in Sect. 4:
Microsoft MixedRealityCompanionKit (See Fig. 14)

Fig. 14. Microsoft spectator-view

https://github.com/Microsoft/MixedRealityCompanionKit/tree/master/SpectatorView
A.2. The main development tool used to develop PC and MR applications Unity3D Game Engine.
https://docs.unity3d.com/Manual/XR.html

References

1. Windows Mixed Reality documentation Available via DIALOG. https://docs.microsoft.com/windows/mixed-reality/. Accessed 20 Apr 2018
2. ARkit Hardware and Software Integration Available via DIALOG. https://developer.apple.com/arkit/. Accessed 20 Apr 2018
3. META2 official community. Available via DIALOG. https://devcenter.metavision.com/support. Accessed 20 Apr 2018
4. A Practical App for Overlaying BIM Models On-Site with Microsoft HoloLens. Available via DIALOG. https://www.engineering.com/ARVR/ArticleID/15095/A-Practical-App-for-Overlaying-BIM-Models-On-Site-with-Microsoft-HoloLens.aspx. Accessed 16 June 2017

5. Oyanagi Construction and Microsoft Japan partner on "Holostruction" project using Microsoft HoloLens. https://news.microsoft.com/apac/2017/05/03/oyanagi-construction-microsoft-japan-partner-holostruction-project-using-microsoft-hololens/. Accessed 3 May 2017
6. https://www.baminternational.com/en/news/bam-takes-it-to-the-next-level-with-the-microsoft-hololens , https://www.baminternational.com/en/news/bam-takes-it-to-the-next-level-with-the-microsoft-hololens. Accessed 22 Feb 2017
7. Ahn, K.J., Kim, S.H., Ko, D.S.: Case study of mixed reality technology in architecture. In: Korean Institute of Information Technology, pp. 110–112 (2017)
8. Doswell, J.T.: Augmented learning: context-aware mobile augmented reality architecture for learning. In: Sixth IEEE International Conference on Advanced Learning Technologies, pp. 1182–1183 (2006)
9. Dunston, P., Wang, X., Billinghurst, M., Hampson, B.: Mixed Reality Benefits For Design Perception, pp. 191–196. NIST SPECIAL PUBLICATION SP (2003)
10. Chen, H., Lee, A.S., Swift, M., Tang, J.C.: 3D collaboration method over HoloLens^{TM} and Skype^{TM} end points. In: ImmersiveME 2015 Proceedings of the 3rd International Workshop on Immersive Media Experiences, pp. 27–30 (2015)
11. Helmut, P.: The global lab: towards a virtual mobility platform for an eco-friendly society. Virtual Reality Soc. Jpn. **14**(2), 163–170 (2009)
12. Hjelseth, S., Morrison, A., Nordby, K.: Design and computer simulated user scenarios: exploring real-time 3D game engines and simulation in the maritime sector. Int. J. Des. **9**(3), 63–75 (2015)
13. Doswell, J.T.: Augmented learning: context-aware mobile augmented reality architecture for learning. In: Proceedings of the Sixth International Conference on Advanced Learning Technologies (2006)
14. Garon, M., Boulet, P.-O., Doiron, J.-P., Beaulieu, L., Lalonde, J.-F.: ¡em¿Real-time high resolution 3D data on the hololens. http://vision.gel.ulaval.ca/-jflalonde/projects/hololens3d/. Accessed July 2016
15. Nakano, J., Osawa, S., Narumi, T., Tanikawa, T., Hirose, M.: Designing a walking tour utilizing on-site virtual time machine. Virtual Reality Soc. Jpn. **22**(2), 241–250 (2017)
16. Takahashi, H., Yoshioka, Y.: Influences of sized and positions of hanging wall on perception of spatial center in virtual reality space. J. Archit. plan AIJ **81**(727), 1905–1915 (2016)

Exploring the Improvement of the Defense Information System

Donghyuk Jo$^{(\boxtimes)}$

Soongsil University, Seoul, South Korea
joe@ssu.ac.kr

Abstract. This study of the information communication technology is intended to suggest Defense Information System (C4I) success model to improve the operational performance of the C4I system and to empirically verify the model. To this end, this study suggested information system quality (system quality, information quality, service quality), perceived usefulness and user satisfaction as antecedent factors affecting the operational performance of the C4I system and then analyzed the effect relationship between the antecedent factors and operational performance. Additionally, this study analyzed the moderating effect of use experiences and IS efficacy. This study is meaningful in that it empirically confirmed the factors affecting the operational performance of the C4I system, and suggested the theoretical foundation and the strategic directions for evaluating and improving the operational performance of the C4I system.

Keywords: Defense information system (C4I) · System quality
Perceived usefulness · User satisfaction · Operational performance
Use experience · IS self-efficacy

1 Introduction

Recently, the 4th Industrial Revolution is expressed as 'the future that is already in front of us', and the latest ICT technology such as robot technology, artificial intelligence, Internet of things has become a new paradigm to decide the future of industry and defense field. The rapid development of information communication technology promotes a transition to the technology and thinking in the information age, and this information revolution based on this information technology and thinking also affect the defense field [1]. The battlefield environment of modern warfare is changing from PCW (Platform Centered Warfare), which mainly relies on the performance of individual weapon system to NCW (Network Centered Warfare), which maximizes the combat mission effect by linking various operational factors and sharing information in real time with each other [2]. To this effect, the Korea's armed forces also invested a large amount of budget in the 2000s in deploying and operating a battlefield management system (Command, Control, Communication, Computer, and Intelligence System: C4I system) to derive the concept of future combat action and implement NCW [3].

To dominate the security environment and gain the relative military power advantages in today's battlefield environment, it is necessary to have tactics of superior

© Springer Nature Switzerland AG 2019
R. Lee (Ed.): ACIT 2018, SCI 788, pp. 62–77, 2019.
https://doi.org/10.1007/978-3-319-98370-7_6

performance and successful operation management strategy for strategic C4I system. Also, due to the continuous development of information communication technology and changes in operational environment, there is the necessity of more advanced battlefield management information system including the improvement of the performance of management information systems and the development of new systems [1]. To successfully manage the C4I system, Korea's armed forces sets budget, schedule, and technical goals in the R & D stage and applies the Earned Value Management System (EVMS) to measure and analyze the performance against the plan. But, there is a lack of system for evaluating overall operational performance in the operation maintenance stage after the deployment for force integration and there are not many studies conducted to evaluate the operational performance of the system due to the specificity of the information system operated by the military [1].

In today's rapidly changing global business environment, companies are actively utilizing the management information system to gain competitive advantages and enhance competitiveness. Therefore, measuring and managing performance throughout the operation of information system is considered to be critical to the business activities of companies. In other words, the information systems have been introduced and used to improve the performance of individuals or organizations in business activities, and the success of such information systems is evaluated by the contribution of information systems to individual business activities and organizational management activities [4]. Delone and McLean [5] reviewed previous studies on IS, and suggested System Quality, Information Quality, Information Use, Individual Impact, and Organizational Impact as important indicators for the success of IS, and suggested IS Success Model by setting the effect relations between indicators. Since then, studies on the success of the information system have been expanded from the success model of the information system based on the performance and quality to Technology Acceptance Model (TAM) and Extended Technology Acceptance Model (Extended TAM), IS Continuance Model, and Social Cognitive Theory- the studies models that that can evaluate the success of information systems at various levels due to the diversification and changes in operational environments [6]. As mentioned above, despite the importance of the studies on the success of the C4I system, which is operated in the field of defense, such studies have been limited to the performance and quality control oriented in the process of C4I system development process. In addition, there is a lack of studies to evaluate the operational performance of the C4I system in the operation and maintenance stage.

Therefore, this study is intended to suggest C4I system success model through a review of the literature about information system and then suggest theoretical foundations and strategic directions for the operational performance improvement of C4I system through the empirical verification of the model.

2 Theoretical Background and Hypothesis

2.1 Battlefield Management Information System

The battlefield environment of the modern warfare is changing from the conventional Platform Centered Warfare (PCW), which greatly relies on the performance of weapon system to Network Centered Warfare (NCW), which maximizes the combat mission effect by linking various operational elements and sharing information in real time through automated information analysis and propagation system [2]. Thus, the role of C4I system (Command, Control, Communication, Computer, and Intelligence System), which is a network-centered battlefield management information system that collects and analyzes quality information from various collection systems and propagates them to the right subjects in a timely manner, is becoming more important [3].

The battlefield management system organically integrates and links various elements such as command, control, communication, computer, and information using an information system to provide accurate battlefield information to military commanders and other members in a timely manner, supporting command and control in real time. More specially, this system visualizes the battlefield situation in conjunction with surveillance and reconnaissance system and hitting system, and enables both upper echelon and lower echelon to recognize the battlefield situation the same and make decision and take actions. Therefore, this system can be defined as 'System of System' [1].

In other words, the C4I system is a complex system in which the distributed weapons are connected by network and communication technology to perform combat mission in the battlefield. Its goals and functions are linked with operational missions so that all elements must be integrated and operated organically without restrictions on process and functions [3]. The development of information and communication technology increases the importance of information superiority, which in turn increases the importance of C4I system day by day. To cope with this change in the times, countries around the world are concentrating on network development and C4I system development. Especially, the importance of the C4I system is emphasized considering our security realities against North Korea and the transfer of wartime operational control [1].

2.2   Information System Success Model

Information system (IS) has been introduced and used to improve the ability of individuals or organizations to perform their missions, and has been evaluated by the contribution of information systems to individual and organizational management activities [4]. However, users of information systems are composed of diverse stakeholders and they have conflicting goals, making it difficult to evaluate information systems [7]. Under these circumstances, Delone and McLean [5] reviewed previous studies on IS and suggested System Quality, Information Quality, Information Use, Individual Impact, and Organizational Impact as important indicators for the success of IS. They set the effect relationship between indictors and suggested IS Success Model. Pitt et al. [7] pointed out that the information system success model of DeLone and McLean [5] overlooked the service aspect of the information system, and suggested the extended information system success model in which the service aspect is applied as information system service factor using SERVQUAL model [4, 6, 8]. Seddon [9] pointed out the limitations and ambiguities of 'use among the measurement variables proposed by DeLone and McLean [5], and suggested 'Perceived Usefulness' in the Technology Acceptance Model [10]. Based on these findings, Seddon [9] suggested an improved information system success model using the Perceived Usefulness as a variable instead of 'Use' in the success model of Lone and McLean (1992) [4]. In addition, DeLone and McLean [5] suggested Updated D & M IS Success Model by adding the service quality factor proposed by Pitt et al. [7] as a new success factor, and changing individual effects and organizational effects to net profit. Therefore, this study will suggest system Quality, Information Quality, Service Quality, Perceived Usefulness, User Satisfaction, Individual Impact and Organizational Impact as success factors based on the previous studies on the success of information system and address these factors, respectively.

2.3 Information Quality System

Today, organizations can improve internal operations efficiency and customer service through effective internal resource management using an information system. The information system quality plays an important role in providing useful values to the organization, such as efficiency of internal operations and improved decision making, by providing timely and accurate information that is beneficial to the organization. Therefore, the information system quality is an important criterion for information system success, and has been evaluated as a sub-dimension of system quality, information quality and service quality [5, 11].

As a sub-dimension of information system quality, the system quality refers to the technological quality of the information system itself that produces and communicates accurate information, which is the technical quality that users perceive during the use of the system [5].

Therefore, the system quality is the degree of the individually perceived performance of a system that processes information [12] and includes access, system functionality, reliability, response time, sophistication, navigation ease and flexibility [6]. Information system uses receive beneficial and accurate information from the information system in a timely manner for the improvement of decision making [11]. Therefore, the information quality can be referred to as the quality of information system process output [5], and includes accuracy, precision, timeliness, sufficiency, understandability, and conciseness [6], And the service quality is the quality of the support that system users receive from information system organizations and IT support personnel [7], and includes responsiveness, accuracy, reliability, and empathy [6].

As such, the information system quality plays an important role in providing values to individuals and organizations, such as efficiency of internal operations and improved decision making, by providing beneficial and accurate information to individuals and organizations in a timely manner [11]. Seddon [9] in his study of information system success demonstrated that system quality and information quality have a significant effect on the perceived usefulness as information system quality. In addition, DeLone and McLean [5] suggested system quality, information quality and service quality as sub-dimensions of information system quality. Subsequent studies empirically demonstrated that system quality has a significant effect on the perceived usefulness [4, 6, 11]. Therefore, this study established hypotheses as follows:

H1a. System quality will have a positive effect on perceived usefulness.
H1b. Information quality will have a positive effect on perceived usefulness.
H1c. Service quality will have a positive effect on perceived usefulness.

Additionally, the information system is also used to achieve specific goals of individuals or organizations. In other words, the value of an information system is realized when it is easy to use, useful, and sustainably implemented [4]. Lone and McLean [5] demonstrated that system quality and information quality have a significant effect on user satisfaction, and Pitt et al. [7] added service quality factor to the information system quality factors of DeLone and McLean [5] and demonstrated that service quality has a significant effect on performance. In addition, the information system quality has been empirically proven to be an important factor in user satisfaction as the attitude of user after the acceptance of information system [4, 6, 11, 13]. Therefore, this study established hypotheses as follows:

H2a. System quality will have a positive effect on user satisfaction.
H2b. Information quality will have a positive effect on user satisfaction.
H2c. Service quality will have a positive effect on user satisfaction.

2.4   Perceived Usefulness and User Satisfaction

The perceived usefulness of information systems is the degree of user's confidence in improving job performance using a system [10] and has been considered a key variable that explains the acceptance. Seddon and Kiew [14] suggested the perceived usefulness as a key determinant of information system success and suggested faster job accomplishment, performance increase, productivity increase, effectiveness increase, job handling ease, and overall system usefulness as the indicators to evaluate the perceived usefulness. In addition, user satisfaction in information systems is the degree of user perceived satisfaction [10], and has been considered as a criterion for measuring information system usefulness [6]. In a study on information system success, Seddon and Kiew [14] suggested user satisfaction as a major determinant of information system success, and suggested satisfying information needs, effectiveness, efficiency, and overall satisfaction as indicators of user satisfaction.

In this way, users receive beneficial and accurate information from the information system in a timely manner for the improvement of decision making and use the information system to improve decision-making [11], and organizations improve the job efficiency of the members by introducing in information system successfully, which leads to performance [6]. Seddon and Kiew [14] and Seddon [9] verified the effect of perceived usefulness on user satisfaction in a study on information system success, and [15] argued that that perceived usefulness in the information system continued use model affects use satisfaction, which leads to the increased continued use intention. In addition, DeLone and McLean [5] argued that information system use and user satisfaction have a significant effect on individual performance. Since then, Pitt et al. [7] and Iivari [16] and Gable et al. [17], empirically proved that the use and user satisfaction have a significant effect on individual performance. Therefore, this study established hypotheses as follows:

H3. Perceived usefulness will have a positive effect on user satisfaction.

H4. Perceived usefulness will have a positive effect on personal performance.

H5. User satisfaction will have a positive effect on personal performance.

2.5   Individual and Organizational Performance

Individual and organizational performance of information system means the degree of the contribution of an information system to individual and organizational success [5] and the individual and organization performance is evaluated by improvements in productivity, quality of decision making, work practices. The purpose of the organization's introduction and operation of the information system is to enhance the performance of individuals by improving the efficiency of the organizational members, and ultimately contribute greatly to the performance of the organization [6], DeLone and McLean [5] and Pitt et al. [7] demonstrated that individual performance has a significant effect on organizational performance in a study of information system success. Therefore, this study established hypotheses as follows:

H6. Personal performance will have a positive effect on organizational performance.

2.6   Use Experience

The user experience in using a new information system affects the continued use of the information system [18]. Users who are experienced in using information systems are more directly informed than those who are less experienced. Also, users who are less experienced in using the information system recognize the new information based on the indirect experience which is not enough to adopt this information system [19]. In other words, as users have different intentions to use technology due to different effects of perception that are formed by direct or indirect experience, the level of user information system use experience affects the user's attitude about using information system [20–23] and performance [24]. Therefore, this study established hypotheses as follows:

H7. Use experience will have a positive effect on IS efficacy.

H8a. Use experience will have a moderating effect on the relationship between perceived usefulness and individual performance.

H8b. Use experience will have a moderating effect on the relationship between user satisfaction and individual performance.

2.7   IS Efficacy

Self-efficacy refers to the confidence in one's own ability to organize various knowledge and skills into context and put them into practice to reach the established performance criteria to solve specific tasks [25]. In the 1990s, with the development of information technology, self-efficacy has become the basis for understanding the behaviors of users through the acceptance and use of information technology along with the development of information technology [26]. Information system self-efficacy refers to the personal perception of the extent to which a user can use the information system to their intended purpose [27]. In other words, the information system efficacy is the ability to perform tasks using an information system, and the ability to use information system successfully helps to accept and use information systems [28]. Therefore, a user with a high level of information system efficacy can recognize and control the functions of the information system better than a user with a low level of information system efficacy, and has a strong intention to use the information system continuously [26, 29, 30]. Therefore, this study established hypotheses as follows:

H9. IS efficacy will have a moderating effect on the relationship between individual performance and organizational performance.

Based on the above hypotheses, study model in this study has been suggested as shown in Fig. 1.

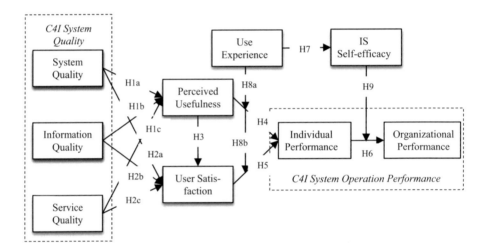

Fig. 1. Research model and hypothesis

3   Research Method

3.1   Samples and Data Collection

This study conducted a survey on users in each group that use C4I system and collected total 300 data in order to verify the suggested study model empirically. After eliminating insincere responses or those with missing values, the final 270 cases were used in the analysis. Characteristics of the sample that participated in the survey of this study are summarized as seen in Table 1.

3.2   Measures

To secure content validity of the measurement tool, this study modified and revised the measurement items from existing literature. First, the sub-dimensions of quality of information system, which consist of system quality, information quality and service quality, were constructed into 4 items each in reference to the studies by DeLone and

Table 1. Sample characteristics

Category and items		Sample size	Ratio (%)
Age (years)	20–29	68	25.2
	30–39	104	38.5
	40–49	93	34.4
	More than 50	5	1.9
Tenure (years)	Less than 5	59	21.9
	5–10	55	20.4
	10–15	112	41.5
	15–20	42	15.6
	More than 20	2	.7
Occupational classification	Combatant	155	57.4
	Information communication	83	30.7
	Logistics support	13	4.8
	Human resource management	2	.7
	Etc.	17	6.3
C4I period of use (years)	Less than 1	3	1.1
	1–3	9	3.3
	3–5	38	14.1
	5–10	99	36.7
	10–15	87	32.2
	15–20	26	9.6
	More than 20	8	3.0

McLean [31], Pitt et al. [7], Seddon [9] and Petter et al. [6], and were measured using 7-point Likert scale (Strongly disagree ∼ Strongly agree). Also, perceived usefulness was constructed into 4 items in reference to the studies by Davis et al. [10], Seddon [9] and Bhattacherjee [15], and was measured using a 7-point Likert scale. User satisfaction was constructed into 4 items in reference to studies by DeLone and McLean [30], Seddon [9], and Bhattacherjee [15], and was measured using a 7-point Likert scale. Individual performance was constructed into 4 items in reference to studies by Pitt et al. [7], Seddon [9], and DeLone and McLean [5], and was measured using a 7-point Likert scale. Organizational performance was constructed into 4 items in reference to studies by Pitt et al. [7], DeLone and McLean [5] and Iivari [16], and was measured using a 7-point Likert scale. IS efficacy was constructed into 4 items in reference to studies by Compeau and Higgins [27], Agarwal and Karahanna [32] and Petter et al. [6], and was measured using a 7-point Likert scale. Lastly, user experience measured the period of C4I system use in reference to the study by Venkatesh et al. [30]. The measurement tool in this study is summarized as Table 2.

Table 2. Measurement model evaluation

Variables		Items	Factor loading	Cronbach's α
C4I system quality	System quality	Ease of system operation	.826	.808
		System speed	.799	
		System credibility	.810	
		System flexibility	.760	
	Information quality	Accuracy of information	.899	.890
		Understandability of information	.881	
		Reliability of information	.813	
		Timeliness of information	.880	
	Service quality	Accessibility of service	.833	.888
		Information connectivity of service	.880	
		Reliability of service	.892	
		Service speed	.861	
Perceived usefulness		Usefulness to user	.924	.939
		Helpful to work	.949	
		Providing helpful information	.928	
		Corresponding to purpose of use	.883	
User satisfaction		Satisfied with use	.925	.935
		Satisfied after use	.919	
		User experience is generally satisfactory	.915	
		Usage is a wise decision	.901	
Operation performance	Individual performance	Job performance	.904	.943
		Convenience	.937	
		Speed	.924	
		Accuracy	.934	
	Organizational performance	Performance and function	.858	.940
		Reliability	.871	
		Operating availability	.857	
		Logistics supportability	.808	
IS self-efficacy		Confident in using	.858	.954
		Has knowledge and technique necessary to use	.894	
		Understands how to use	.861	
		Proficient at using	.907	

4   Analysis and Results

4.1   Correlations Among Variables

To verify the correlations among variables, correlation analysis was performed. As a result, the correlations among variables corresponded to the study model suggested in this study, suggesting that the in-depth analysis of the study model was valid. The analysis of correlations among variables in this study is summarized as Table 3.

Table 3. Correlations among variables

Variable	1	2	3	4	5	6	7
1. System quality	1.000						
2. Information quality	.558**	1.000					
3. Service quality	.344**	.613**	1.000				
4. Perceived usefulness	.378**	.615**	.561**	1.000			
5. User satisfaction	.625**	.718**	.575**	.664**	1.000		
6. Individual performance	.375**	.526**	.433**	.619**	.606**	1.000	
7. Organizational performance	.452**	.406**	.384**	.426**	.484**	.488**	1.000

4.2   Main Effect Model

The results of Main effect model's are as follows. In terms of the effects of quality of information system on perceived usefulness, information quality (t = 6.151, p = .000) and service quality (t = 5.033, p = .000) had significant effects on perceived usefulness, but system quality (t = .901, p = .368) did not have a significant effect on perceived usefulness, rejecting hypothesis H1a and adopting hypotheses H1b and H1c (Table 4).

Table 4. Relationship between information system quality and perceived usefulness

Dependent variable	Independent variable	β	t
Perceived usefulness	System quality	.050	.901
	Information quality	.407	6.151***
	Service quality	.294	5.033***
$R^2 = .434$, F = 67.942, P = .000			
***<.001			

In terms of the effects of quality of information system on user satisfaction, system quality (t = 7.108, p = .000), information quality (t = 7.461, p = .000), and service quality (t = 4.458, p = .000) all had significant effects on user satisfaction, adopting hypotheses H2a, H2b, and H2c (Table 5).

Table 5. Relationship between information system quality and user satisfaction

Dependent variable	Independent variable	β	t
User satisfaction	Perceived usefulness	.664	14.549[***]
$R^2 = .441$, F = 211.663, P = .000			

***<.001

In terms of the effects of perceived usefulness on user satisfaction, perceived usefulness (t = 14.549, p = .000) had a significant effect on user satisfaction, adopting hypothesis H3 (Table 6).

Table 6. Relationship between perceived usefulness and user satisfaction

Dependent variable	Independent variable	β	t
User satisfaction	Perceived usefulness	.664	14.549[***]
$R^2 = .441$, F = 211.663, P = .000			

***<.001

In terms of the effects of perceived usefulness on individual performance, perceived usefulness (t = 12.903, p = .000) had a significant effect on individual performance, adopting hypothesis H4 (Table 7).

Table 7. Relationship between perceived usefulness and user satisfaction

Dependent variable	Independent variable	β	t
Individual performance	Perceived usefulness	.619	12.903[***]
$R^2 = .383$, F = 166.482, P = .000			

***<.001

In terms of the effects of user satisfaction on individual performance, user satisfaction (t = 12.488, p = .000) had a significant effect on individual performance, adopting hypothesis H5 (Table 8).

Table 8. Relationship between user satisfaction and individual performance

Dependent variable	Independent variable	β	t
Individual performance	User satisfaction	.606	12.488[***]
$R^2 = .368$, F = 155.942, P = .000			

***<.001

In terms of the effects of individual performance on organizational performance, individual performance (t = 9.153, p = .000) had a significant effect on organizational performance, adopting hypothesis H6 (Table 9).

Table 9. Relationship between individual performance and organizational performance

Dependent variable	Independent variable	β	t
Organizational performance	Individual performance	.488	9.153***
R^2 = .238, F = 83.786, P = .000			

***<.001

In terms of the effects of user experience on IS efficacy, user experience (t = 2.661, p = .008) had a significant effect on IS efficacy, adopting hypothesis H7 (Table 10).

Table 10. Relationship between use experience and IS self-efficacy

Dependent variable	Independent variable	β	t
IS self-efficacy	Use experience	.160	2.661**
R^2 = .026, F = 7.079, P = .000			

**p < .01

4.3 Moderating Effect Model

The results of Moderating effect model's are as follows. In terms of the moderating effect of user experience on the relationship between perceived usefulness and individual performance, user experience (t = 2.603, p = .010) had significant effects on all, adopting hypothesis H8a. Also, in terms of the moderating effect of user experience on the relationship between user satisfaction and individual performance, user experience (t = 3.197, p = .002) had significant effects, adopting hypothesis H8b (Table 11).

Table 11. Moderating effects of use experience

Dependent variable	Moderating variable	Independent variable	β	t
Individual performance	Use experience	Perceived usefulness	.754	2.603**a
Individual performance	Use experience	User satisfaction	.955	3.197**b

*p < .05, **p < .01, ***<.001
[a]Perceived Usefulness *Use Experience
[b]User Satisfaction *Use Experience

In terms of the moderating effect of IS efficacy on the relationship between individual performance and organizational performance, IS efficacy (t = 3.818, p = .000) had significant effects, adopting hypothesis H9 (Table 12).

Table 12. Moderating effects of use experience

Dependent variable	Moderating variable	Independent variable	β	t
Organizational performance	IS self-efficacy	Individual performance	1.320	3.818^{***a}

*p < .05, **p < .01, ***<.001
[a]Individual Performance *IS self-efficacy

5 Conclusions

To dominate the security environment and gain the relative military power advantages in today's battlefield environment, it is necessary to have tactics of superior performance and successful operation management strategy for strategic C4I system. Also, due to the continuous development of information communication technology and changes in operational environment, there is the necessity of more advanced battlefield management information system including the improvement of the performance of management information systems and the development of new systems. [32–34] To successfully manage the C4I system, Korea's armed forces sets budget, schedule, and technical goals in the R & D stage and applies the Earned Value Management System (EVMS) to measure and analyze the performance against the plan. But, there is a lack of system for evaluating overall operational performance in the operation maintenance stage after the deployment for force integration and there are not many studies conducted to evaluate the operational performance of the system due to the specificity of the information system operated by the military [1].

Therefore, this study is determine the factors affecting the operational performance of C4I system, reviewed information system success models suggested in previous studies, and based on the finding, suggested the information system quality of C4I system (system quality, information quality, service quality), perceived usefulness, user satisfaction and operational performance (personal performance, organizational performance) as the effect factors affecting the operational performance, and empirically verified the causal relationship between factors. The conclusions have been derived as follows:

First, information quality and service quality, which are the sub-dimensions of information system quality, have a positive effect on perceived usefulness; and system quality, information quality and service quality have positive effects on user satisfaction. In other words, the C4I system provides beneficial values such as efficiency of internal operation and improved decision making by providing beneficial and accurate information to users in a timely manner [11], and continuously provide the functions and performance needed for system users to perform tasks efficiently, affecting a positive effect on user satisfaction [4]. However, the system quality does not have a significant effect on perceived usefulness, which can be interpreted as the difference in perception depending on the usage environment and ability of the system users affects the attitude of the user.

Second, perceived usefulness as an attitude of information system users has a positive effect on user satisfaction; and perceived usefulness and user satisfaction have a positive effect on individual performance. This results suggest that if users of C4I system perceive the usefulness of system such as helps to job performance and job effectiveness, this will increase user satisfaction on system functions and outputs, thereby increasing user efficacy [6].

Third, the individual performance of information system users has a positive effect on organizational performance. In other words, the C4I system improves the job efficiency of the users, enhances the individual performance, and ultimately contributes to the organizational performance [6].

Fourth, the experience of using an information system has a positive effect on information system efficacy. In other words, the user acquires more information about the information system through the experience of using the information system, and therefore they recognize that they use system successfully as they have more experience of using CI4 system [28].

Fifth, the use experience has a moderating effect on the relationship between perceived usefulness, user satisfaction and individual performance. In other words, the CI4 system user perception on system varies developing on the level of the experience of using the system, which has a significant effect on the attitude of the user [21].

Finally, IS efficacy has a moderating effect on the relationship between individual performance and organizational performance. In other words, the ability of the user to successfully use the information allow the users to recognize and control the functions of CI4 system better, positively affecting organizational performance [25].

Taken together, user's perceived usefulness of the system through the excellent information system quality of the C4I system and the increase in satisfaction improve the performance of the individual users, leading to the improvement of the organizational performance. Therefore, to improve the operational performance of the C4I system, it is necessary to make various efforts to improve the performance and quality of the system in the R & D and performance improvement process of the C4I system. In addition, as users' IS efficacy affects operational performance and operational performance of C4I system, it is also important for users of C4I system to have confidence in using the system by securing a training system for initial system users.

This study suggested a C4I system success model and empirically verified the model, proving its theoretical extension. Existing studies of information system success have been focused on the information systems related to management activities in the private sector, and studies on the success of information systems in the defense sector have been limited. Therefore, this study is meaningful in that it suggested a success model of defense information system and verified it empirically, extending the information success model to the defense field. In addition, this study empirically confirmed the factors affecting the operational performance of the C4I system, and suggested the theoretical foundation and the strategic directions for evaluating and improving the operational performance of the C4I system.

References

1. Ministry of National Defense: 2015 Defense White Paper, Korea (2015)
2. Alberts, D.S., Garstka, J.J., Stein, F.P.: Network Centric Warfare: Developing and Leveraging Information Superiority. Assistant Secretary of Defense (C3I/Command Control Research Program), Washington DC (2000)
3. Cha, H.J., Kim, J.M., Ryou, H.B., Jeong, H.Y.: A study on the improvement of interoperability in Rok C4I system for future warfare. In: Yeo, S.S., Pan, Y., Lee, Y., Chang, H. (eds.) Computer Science and its Applications, pp. 805–812. Springer, Dordrecht (2012)
4. Petter, S., DeLone, W., McLean, E.: Measuring information systems success: models, dimensions, measures, and interrelationships. Eur. J. Inf. Syst. **17**(3), 236–263 (2008)
5. Delone, W.H., McLean, E.R.: The DeLone and McLean model of information systems success: a ten-year update. J. Manag. Inf. Syst. **19**(4), 9–30 (2003)
6. Petter, S., DeLone, W., McLean, E.R.: Information systems success: the quest for the independent variables. J. Manag. Inf. Syst. **29**(4), 7–62 (2013)
7. Pitt, L.F., Watson, R.T., Kavan, C.B.: Service quality: a measure of information systems effectiveness. MIS Q. **19**(2), 173–187 (1995)
8. Parasuraman, A., Zeithaml, V.A., Berry, L.L.: Servqual: a multiple-item scale for measuring consumer Perc. J. Retail. **64**(1), 12–40 (1988)
9. Seddon, P.B.: A respecification and extension of the DeLone and McLean model of IS success. Inf. Syst. Res. **8**(3), 240–253 (1997)
10. Davis, F.D., Bagozzi, R.P., Warshaw, P.R.: User acceptance of computer technology: a comparison of two theoretical models. Manag. Sci. **35**(8), 982–1003 (1989)
11. Gorla, N., Somers, T.M., Wong, B.: Organizational impact of system quality, information quality, and service quality. J. Strateg. Inf. Syst. **19**(3), 207–228 (2010)
12. Freeze, R.D., Alshare, K.A., Lane, P.L., Wen, H.J.: IS success model in e-learning context based on students' perceptions. J. Inf. Syst. Educ. **21**(2), 173–184 (2010)
13. Lin, H.F.: Measuring online learning systems success: applying the updated DeLone and McLean model. Cyberpsychol. Behav. **10**(6), 817–820 (2007)
14. Sedon, P., Kiew, M.: A partial test and development of the DeLone and McLean model of IS success. In: DeGross, J.I., Huff, S.L., Munro, M.C. (eds.) Proceeding of the Fifteenth International Conference on Information Systems, pp. 99–110 (1994)
15. Bhattacherjee, A.: Understanding information systems continuance: an expectation-confirmation model. MIS Q. **25**(3), 351–370 (2001)
16. Iivari, J.: An empirical test of the DeLone-McLean model of information system success. ACM SIGMIS Database DATABASE Adv. Inf. Syst. **36**(2), 8–27 (2005)
17. Gable, G.G., Sedera, D., Chan, T.: Re-conceptualizing information system success: the IS-impact measurement model. J. Assoc. Inf. Syst. **9**(7), 377–408 (2008)
18. Jayawardhena, C.: Personal values' influence on e-shopping attitude and behaviour. Internet Res. **14**(2), 127–138 (2004)
19. Karahanna, E., Straub, D.W., Chervany, N.L.: Information technology adoption across time: a cross-sectional comparison of pre-adoption and post-adoption beliefs. MIS Q. **23**(2), 183–213 (1999)
20. Lim, E.T., Pan, S.L., Tan, C.W.: Managing user acceptance towards enterprise resource planning (ERP) systems–understanding the dissonance between user expectations and managerial policies. Eur. J. Inf. Syst. **14**(2), 135–149 (2005)
21. Morris, M.G., Venkatesh, V.: Age differences in technology adoption decisions: implications for a changing work force. Pers. Psychol. **53**(2), 375–403 (2000)

22. Nevo, D., Chan, Y.E.: A temporal approach to expectations and desires from knowledge management systems. Decis. Support Syst. **44**(1), 298–312 (2007)
23. Venkatesh, V., Davis, F.D.: A theoretical extension of the technology acceptance model: four longitudinal field studies. Manag. Sci. **46**(2), 186–204 (2000)
24. Staples, D.S., Wong, I., Seddon, P.B.: Having expectations of information systems benefits that match received benefits: does it really matter? Inf. Manag. **40**(2), 115–131 (2002)
25. Bandura, A.: Self-efficacy: toward a unifying theory of behavioral change. Psychol. Rev. **84**(2), 191 (1977)
26. Bhattacherjee, A., Perols, J., Sanford, C.: Information technology continuance: a theoretic extension and empirical test. J. Comput. Inf. Syst. **49**(1), 17–26 (2008)
27. Compeau, D.R., Higgins, C.A.: Computer self-efficacy: development of a measure and initial test. MIS Q. **19**(2), 189–211 (1995)
28. Compeau, D., Higgins, C.A., Huff, S.: Social cognitive theory and individual reactions to computing technology: a longitudinal study. MIS Q. **23**(2), 145–158 (1999)
29. Yang, K.: Consumer technology traits in determining mobile shopping adoption: An application of the extended theory of planned behavior. J. Retail. Consum. Serv. **19**(5), 484–491 (2012)
30. Venkatesh, V., Morris, M.G., Davis, G.B., Davis, F.D.: User acceptance of information technology: toward a unified view. MIS Q. **27**(3), 425–478 (2003)
31. DeLone, W.H., McLean, E.R.: Information systems success: the quest for the dependent variable. Inf. Syst. Res. **3**(1), 60–95 (1992)
32. Agarwal, R., Karahanna, E.: Time flies when you're having fun: cognitive absorption and beliefs about information technology usage. MIS Q. **24**(4), 665–694 (2000)
33. Yoo, C.S., Jeong, E.J.: A study on the improvement for management efficiency about the approach of applying IT to our defense area. Q. J. Def. Policy Stud. **92**, 67–99 (2011)
34. Kim, C.M., Kim, I.: A study of influencing factors upon using C4i systems: the perspective of mediating variables in a structured model. Asia Pac. J. Inf. Syst. **19**(2), 73–94 (2009)

A Modern Solution for Identifying, Monitoring, and Selecting Configurations for SSL/TLS Deployment

Lamya Alqaydi$^{(\boxtimes)}$, Chan Yeob Yeun$^{(\boxtimes)}$, and Ernesto Damiani$^{(\boxtimes)}$

ECE Department, Khalifa University of Science and Technology,
Abu Dhabi, UAE
lalqaydi@gmail.com, cyeun@ku.ac.ae,
Ernesto.damiani@kustar.ac.ae

Abstract. Some of the well-known vulnerabilities like DROWN, POODLE, and Heartbleed affect a subset of all possible configurations of protocols and cipher-suites in SSL/TLS protocol. Recently, new vulnerabilities are also frequently discovered and could be used to mount attacks on systems whose configurations are not updated in time or were misconfigured from the start. Thus, we provide an overview of the landscape of vulnerabilities relating to SSL/TLS protocol versions with estimated risk levels. Selecting the best configuration for a given use-case is a time-consuming task and testing a given configuration of a server for all known vulnerabilities is also difficult. Thus, there is a great motivation to create a new tool that abstracts the tedious parts of this process. Our new software solution can automatically scan and rate the configuration of servers and help in selecting suitable ones. The goal is to simplify testing and evaluation of server-side configurations of SSL/TLS and ciphersuites for the community and thus the software is released as an open source.

Keywords: TLS · SSL · Privacy · Security · TLS handshake
Renegotiaition attack · BEAST attack · CRIME and BREACH attack
Heartbleed attack

1   Introduction

There are compatibility limitations and different possible ciphersuite combinations used in implementations of SSL/TLS protocols. New vulnerabilities related to SSL/TLS protocols are frequently discovered and could be used to mount attacks on systems whose configurations are not updated in time or were misconfigured from the start.

Selecting a good configuration and continuously testing it for vulnerabilities is outside the expertise of most developers and web administrators. When it comes to SSL/TLS protocols, this is exacerbated by the differing levels of support for protocol versions and ciphersuites among client devices and servers [1]. Most clients and servers support a subset of the two versions of SSL (v2 and v3) and three earlier versions of TLS (v1.0, v1.1, and v1.2). TLS v1.2 is the most secure but not all clients or servers support it. Some modern browsers have started adding support for the upcoming TLS v1.3 as well.

© Springer Nature Switzerland AG 2019
R. Lee (Ed.): ACIT 2018, SCI 788, pp. 78–88, 2019.
https://doi.org/10.1007/978-3-319-98370-7_7

In addition, some of the well-known vulnerabilities like DROWN, POODLE, and Heartbleed affect a subset of all possible configurations of protocols and ciphersuites. Selecting the optimal configuration is a time-consuming task and testing a given configuration of a server for all known vulnerabilities is even more so. Even setting up parameters for widely-used implementations can be difficult. For example, the libcurl library used by all of Amazon's API products has a configuration variable called CURLOPT_SSL_VERIFYHOST that needs to be set to '2' rather than 'true' or '1' in order to have certificate validation.

Given these facts, the problem selected for this research paper is reducing the difficulty of knowing how strong SSL/TLS configurations are and how to improve them.

2 Literature Review

Historical Overview
At the start of the Internet, the used communication protocols, such as HTTP and FTP, were designed with no concern for security and relied on honest behavior. However, as the possibility of e-commerce applications began to attract attention, efforts were done to secure communication links. Researchers and developers created authentication methods and added optional encryption to the transport layer of the OSI model, in the form of the SSL/TLS protocol. Authentication results in sensitive data being sent only to the correct recipients and the inclusion of encryption means that an attacker who manages to obtain the encrypted messages would not gain information or succeed in changing the content of messages easily.

The Secure Sockets Layer (SSL) was first invented by Netscape in 1994 and proposed as a solution for securing communications at the transport layer of the OSI model for purposes for e-commerce [2]. In 1999, the SSL protocol became the responsibility of the IETF (Internet Engineering Task Force) and it was renamed to Transport Layer Security protocol TLS [3]. Six versions of the protocol were released to the public so far, SSLv2, SSLv3, TLSv1.0, TLSv1.1, TLSv1.2, and TLSv1.3, which is still in development.

SSLv3 was released shortly after SSLv2 in order to fix structural security flaws in SSLv2 and add support for certificate chains, renegotiation, compression, more cryptographic algorithms, and anonymous suites [4]. TLSv1.0 is a minor update to SSLv3 that improved the security and branding of the protocol and is actually encoded as v3.1 [3]. Protocol versions TLSv1.1 and TLSv1.2 were released more recently in 2006 and 2008 and they both contain major changes to protect against known or suspected vulnerabilities at the time [5–7].

The SSL/TLS Protocol and Ciphersuites
The SSL/TLS layer is flexible in that it can be configured to use different cryptographic algorithms and support subsets of protocol versions and extensions. In this section, the working of the SSL/TLS protocol is summarized and the role of cryptographic algorithms and ciphersuites in protocol configurations is specified [8, 9].

How SSL/TLS Work

The role of the SSL/TLS protocol is to authenticate the server/client pairs and to provide a public key exchange mechanism. The evolution of the design of the SSL/TLS protocol makes sense considering that the priorities of the protocol design are to enable cryptographically secure communication, be interoperable across platforms and developers, allow for extensions, and have good performance.

The SSL/TLS protocol consists of four sub-protocols and it also has extensions starting from TLSv1 (IETF, 2006). The two main extensions are Session Tickets and Signature Algorithms while the component protocols are surveyed next:

Handshake Protocol

The Handshake protocol initiates and starts the communication between the client and the server. It involves the negotiation of sessions, SSL/TLS protocol versions, ciphersuites, certificate chain exchanges and any needed authentication. It follows the process shown in Fig. 1.

Since this is the process which is most relevant to the goals of this paper, here is a summary of how the SSL/TLS handshake happen:

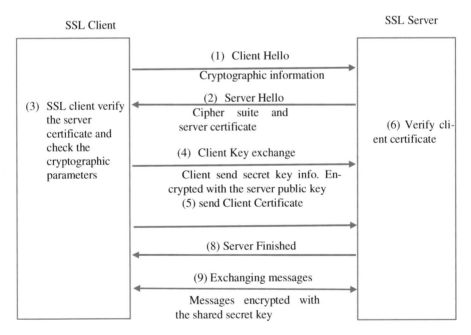

Fig. 1. Handshake between the SSL client and the SSL server

Step 1: The SSL/TLS client sends "Client Hello" message that includes the information about TLS/SSL version and the cipher suites that are supported by the client.
Step 2: The SSL/TLS server responds with a "server Hello" that has the cipher suite selected by the server. It also contains session ID in addition to that the server sends its digital certificate to the client.

Step 3: The SSL/TLS client verify the servers' digital certificate by the server public key.

Step 4: The SSL/TLS client then will send a randomly generated byte that will be used to find the secret key that will be used to encrypt the coming conversations. This randomly generated byte will be encrypted by the public key of the server and the server can decrypt it using its private key.

Step 5: if the SSL/TLS server requested from the client a certificate request. Step 4 will be done in addition to the clients' digital certificate that will be sent to the server. And in this stage the server can be able to verify the identity of the client.

Step 6: The SSL/TLS server will verify the client certificate.

Step 7: The SSL/TLS client sends "client finished" that is encrypted by the previously generated secret key. This will be an indication to the server that the client finished its part in the handshake.

Step 8: The SSL/TLS server sends "server finished" message to indicate that the server handshake is finished as well.

Step 9: Throughout the session that was initiated once the handshake started both the server and client were able to send and receive messages that uses symmetric encryption mechanism [15–17].

Record

In this layer, messages have a header and a hash value. The payload can be compressed, and the messages can encapsulate verification of payload integrity and length as well as the protocol parameters.

Cipher Spec

This is the start of secure communication as the cryptographic suite which was agreed on by the server/client pair is used to encrypt all messages.

Alert

Any errors or warning that occur are reported here in the form of Alert Description and Severity Level.

Ciphersuites and Configurations

Together, the algorithms used for a connection are called a ciphersuite. The set used for communication between a server/client pair is determined dynamically when the connection is started as was explained in the previous subsection. The choice of a ciphersuite for use in SSL/TLS protocol specifies the authentication method, key exchange, hash function, and MAC algorithm among others. The SSL/TLS protocols can generally use RSA, EC, AES-CBC, AES-GCM, RC4, 3DES, MD5, SHA1, MACs, Signatures, and PRFs as cryptographic primitives.

We have currently some 364 different ciphersuites identified in the testing database. These ciphersuites differ in their levels of security, compatibility with versions of SSL/TLS, and also in their computational requirements.

There exist several naming schemes for ciphersuites. We use the IANA names which are constructed from the cryptographic primitives used in the ciphersuite as shown in Fig. 2.

Fig. 2. IANA naming scheme for ciphersuites. Adapted from [19]

Vulnerabilities

The vulnerabilities discussed here should be taken in the context of SSL/TLS configurations and not necessarily other secure-communication protocols which have some analogous vulnerabilities that are outside the scope.

Vulnerabilities in secure protocols result from an attacker finding the secret key via some computational brute force search combined with weaknesses in the design of the implementation or cryptographic primitive or from bugs in the implementations' code. As is often said, cryptography is more often bypassed rather than broken [10, 11].

In general, attacks can be divided into two categories: passive and active. Passive attacks involve spying on connections to learn identities and cryptographic capabilities of clients and servers. Active attacks, on the other hand, change the messages being sent and potentially have the attacker impersonating either the server or the client [18–20].

Table 1. Classification of vulnerabilities affecting the SSL/TLS protocols [12, 13, 21–24, 26, 27, 31–33]

Type of vulnerability	Risk Levels	Notable examples and causes
Specification Flaws	High	SSLv2 is fundamentally insecure
Implementation Flaws	High	Memory management bugs (Heartbleed), state ma- chine faults, SMACK (Benjamin Beurdouche et al., 2015) (Benjamin Beurdouche, 2015)
Weak Hash Functions	Moderate	MD5 is susceptible to collision attacks. SHA-1 and DSA are weak
Short Cryptographic Key lengths	High to Moderate	Often a result of exports restrictions. Recent developments allow for breaking some encryption schemes by brute force
Cross-protocol attacks	Moderate	Servers that try to be compatible with old protocols can be fooled to use the weaker schemes. FREAK at- tacks impersonate servers using 512bits RSA keys. (N. Mavrogiannopoulos, 2012)
Padding Oracle/CBC attacks	High	Lucky 13, POODLE and BEAST
Compression analysis attacks	Moderate	CRIME, TIME and BREACH
RC4 statistical biases attacks	High	Break encryption of WEP, recover secrets (Christina Garman, 2015)
Randomness generation attacks	Moderate	Bad randomness generators or reuse of generated numbers. Sony reused a nonce to sign PS3 firmware
Certificate authority attacks	Low	French IGC/A certificate authority was used to sign fake Google domains

There are many vulnerabilities affecting the SSL/TLS protocol implementations so far. This is not unexpected because as more valuable services and data are offered online, the incentives are ever higher for attackers. In 2014 alone, new vulnerabilities were discovered in OpenSSL [30] (Heartbleed), SChannel (remote code execution on Windows servers), NSS (Signature forgery in Firefox), and GnuTLS [28] (impersonation of any server).

Considerable research has been done to create a listing of all known vulnerabilities and which configurations they affect as part of this paper. The data will be released through a GitHub repository for the wider community to use and contribute to. An overview of vulnerability types and risk levels is shown in Table 1.

3   Design of the System

Determination of Requirements

Build an easy to use tool for testing the SSL/TLS configuration of any server or endpoint connected to the Internet with good explanations of the results. Make the tool along with all needed data available for further development and maintenance by the wider security community.

For the first requirement, several sub-requirements were specified. First, the tool must accept either an IP address or a domain name to identify the target of the test. Second the results must be shown in color coded tables and summarized with graphics such as star ratings and notes which describe the issues that must be fixed and danger level of any vulnerability. Finally, the website version of the tool must offer website monitoring service and signing up to receive alerts for users.

As for the second requirement, several sub-requirements were also chosen. The programming languages and technology stacks used for the development of the tool and the storage of the data must be in active use and likely to have developers in the foreseeable future so that the tools do not become obsolete or too difficult to maintain. In addition, the created software is to be shared on Github and developed in a Docker container environment so that deployment of the system is automated and issues with versions of libraries or operating systems do not affect the product [14]. The last requirement is that of having clear documentation and fair licensing to lower barriers of entries for new developers and prevent the abuse of the development efforts of the community.

Architecture of the System

There are three main components to the paper described in this paper. The first component is a backend server set up to scan and rate any network-connected host for supported SSL/TLS protocol versions and ciphersuites. The second is an open source repository of protocols, ciphersuites, and the vulnerabilities that affect each of them. This repository is used to generate, and update databases used for testing the servers' configurations. Finally, a web application, which connects to the backend, makes it easy for users to test their servers and choose an appropriate configuration. A Docker container is provided to allow for secure testing of internal sites using self-hosted versions of our software.

At the start, a user can go to the hosted website, tlssuites.com, and input an IP address or a domain name with an associated server they want to have scanned. Once the user agrees to usage terms, a request is sent from the client side of the application to the backend where the request is validated and processed. If the server had not been scanned recently, its SSL/TLS configuration is determined experimentally by trying to connect with each protocol version and then each possible ciphersuite. In addition, the choice the server makes for the ciphersuite for emulated user agents is noted. Once the data is gathered, the server configuration is evaluated based on the researched criteria with regards to protocol support and vulnerabilities. The resulting evaluation is optionally stored, and the results are sent back to the client side to be displayed in an understandable manner for the user.

Evaluation of Configurations
The goal of the evaluation of SSL/TLS configurations is coming up with simple to understand explanations of what potential vulnerabilities exist in the configuration and whether anything should or could be done about them. With this in mind, we have developed a rating system of the strength of configurations and individual components.

Cryptographic primitives are rated according to standard methods. Key exchange parameters such as key length and algorithm are given ratings based on number of bits of security they provide which is a measure of the computational effort that would be required to break the encryption. Any key exchange setup with no support for forward secrecy is automatically rated as insecure as any breach of the key could be used to decode all previous messages which might have been archived by an attacker. This rating applies to RSA which should not be used for encryption but is a valid choice for authentication. Table 2 shows the recommended key lengths adapted from the BSI – Technical Guideline TR-02102-1 released in January of 2018.

Table 2. Recommended key lengths for use until the year 2022. Adapted from BIS - TR-02102-1.

Block cipher	MAC	RSA	DHE	ECDHE	ECDSA
128	128	2000	2000	250	250

Support for Selecting Configurations
The web application includes an interactive decision tree that guides users in determining optimal configurations for their servers based on their answers to Yes/No and checkbox questions about the requirements of their use case. The decision tree takes into consideration the recommendations of trusted sources such as The Operations Security (OpSec) team at Mozilla (MozillaWiki, 2018) and CELLOS consortium (Cellos-consortium.org, 2018). It excludes any configurations known to be vulnerable as updated in the community database. The recommendations include the following considerations.

1. Key Sizes and Algorithms

For the recommended minimum of 128 bits of s security, a key algorithm such as 256-bit ECDSA should be used. If RSA is to be used for compatibility reasons, key sizes > 3,072 k bits should be used given the hardware applicability should be used and internet is must be the perfect choice.

2. Certificate

For the creation of appropriate certificates, it is important to choose a good CA. Let's Encrypt is a good choice which is completely free and comes with certbot which is a command line utility used for configuring different types of computers to support safe configurations of SSL/TLS when using popular tech stacks with NGINX or Apache [29].

Keys and certificates should not be shared among servers and especially not between servers where one accepts SSLv3. For reasons of efficiency it is desirable to use the minimum number of certificates required while still providing the full chain of authority.

3. Signature Algorithm

MD5 should never be used for hashing. SHA1 may only be used to support Windows XP machines without the update SP3 that are of extreme importance that not supporting them is not an option. Even in this case it should be understood that SHA1 has only 61 strength and measures should be taken to prevent its use with non-Windows XP machines. Who is the need to support old Windows XP machines, ensure that you use SHA256 or better.

4. Protocol Versions

When choosing what subset of protocol versions to accept, limitations are imposed by compatibly needs with old devices and minimum-security standards relevant to the company such as PCI Data Security Standard (PCI DSS) or Federal Information Processing Standard (FIPS).

The most secure protocol version is TLSv1.2 but it's a good idea to add versions 1.1 and 1.0 as they do not add significant security risks but allow for secure interaction all browsers more modern than IE6 [34–36].

5. Ciphersuites

The most important often ignored configuration parameters for ciphersuites is the specification and enforcing of the ordering of ciphersuites in terms of preferences for each protocol version or client. The message encoding mode offered for SSLv3, if it needs to be accepted, and TLS versions should be make explicit. The parameters should be set with the knowledge that while RC4 is susceptible to active attackers, it is safer than CBC when it comes to SSLv3.

Servers generally should include support for around 10 or more ciphersuites for compatibility with most devices in use. The choice of ciphersuite should be made to have at least 128 bits of security with authenticated suites that use GCM whenever possible. The NSA Suite B cryptography standard recommends using only GCM suites.

RSA should not be used for key exchanges because it does not have forward secrecy. ECDHE with the curves secp256r1 or secp384r1 is more performant than DHE but either could be used to get the wanted number of bits of security.

Note that the performance hit of using the SSL/TLS protocol is not very significant as Google's measurements show increases of less than 1% in CPU utilization and less than 10 KB of memory per connection and less than 2% of overhead. The following figure shows some performance parameters for the use of different key exchange algorithms in SSL/TLS protocol (Fig. 3).

Fig. 3. A snapshot of the frontend of the system that is the user interface used to test different targets for the security of SSL/TLS used.

Monitoring Service. Website administrators would find the security-monitoring feature of this paper especially useful. They are able to subscribe to have their servers tested daily and after each update to the vulnerability database. They could also register to receive emails and/or text messages to alert them of any changes to the security of their servers' configurations.

Community Support. The database of ciphersuites, protocols, and vulnerabilities used by the backend is dynamically generated from JSON records in a public repository. From the start, it integrates the naming schemes of SSL and IANA [19]. The records include data regarding the support of different SSL/TLS protocol versions in modern browsers, security status of different ciphersuites, and any known vulnerabilities with links to references and additional details. The repository is where community members can contribute updates in the form of pull requests that are integrated after a review process.

4 Conclusion

The performance of the system has been validated by testing its ability to identify the configuration of Khalifa university servers. It has managed to identify the correct configuration and evaluate it appropriately, discovering possibilities along the way.

There are several improvements to the system considered for future development. It would be beneficial to have the testing backend simulate handshakes with different user agent to test the enforcement of ciphersuite and protocol version preferences. In addition, the ability to check the certificates chain and investigate the authenticity of each one taking notes of those revoked or going to expire soon and include this feature in the monitoring system for registered users.

Finally, we have some plans for additional utilities related to SSL/TLS configurations. For example, the output given in the IANA naming scheme but could be translated to other naming schemes. It would also be useful to show the strength of encryption and authentication specified in ciphersuites in terms of bits of security.

We set out to empower application developers and security researchers to identify security issues related to SSL/TLS protocols in advance and act to prevent further issues from arising.

There is still some work to be done for this paper to have the usability and power that would make it indispensable. We have plans for future of this paper and once it is released as open source we would focus on building the community who would maintain and contribute to the development.

References

1. Ristić, I.: Bulletproof SSL and TLS, 1st edn. Feisty Duck (2014)
2. Hickman, K.: The SSL Protocol. Netscape Communications Corp, 9 February 1995
3. Dierks, T., Allen, C.: The TLS Protocol Version 1.0. RFC 2246, January 1999
4. Freier, A., Karlton, P., Kocher, P.: The Secure Sockets Layer (SSL) Proto- col Version 3.0. From RFC 6101 (Historic), Internet Engineering Task Force, August 2011. http://www.ietf.org/rfc/rfc6101.txt
5. IETF: Protocol Version 1.1. RFC 4346, April 2006
6. IETF: The Transport Layer Security (TLS) Protocol Version 1.2, August 2008. http://www.ietf.org/rfc/rfc5246.txt
7. Freier, A., Karlton, P., Kocher, P.: The SSL 3.0 Protocol. Netscape Communications Corp., 18 November 1996
8. AWS: Alexa Top Sites - Up-to-date lists of the top sites on the web, February 2018. https://aws.amazon.com/alexa-top-sites/
9. Bash: GNU Project - Free Software Foundation, January 2018. https://www.gnu.org/software/bash/
10. Beurdouche, B., et al.: SMACK: State Machine AttaCKs, March 2015. https://www.smacktls.com/
11. Beurdouche, B., Bhargavan, K., Delignat-Lavaud, A., Fournet, C., Kohlweiss, M., Pironti, A., Strub, P.Y., Zinzindohoue, J.K.: A messy state of the union: taming the composite state machines of TLS. In: 2015 IEEE Symposium on Security and Privacy, SP 2015, San Jose, CA, USA, pp. 535–552 (2015)
12. Cellos-consortium.org: Cryptographic protocol Evaluation toward Long-Lived Outstanding Security Consortium (CELLOS) (2018). https://www.cellos-consortium.org/index.php. Accessed 11 Feb 2018
13. Christina Garman, K.G.: Attacks only get better: password recovery attacks against RC4 in TLS. In: 24th USENIX Security Symposium, Washington, D.C., USA, pp. 113–128 (2015)
14. Docker Documentation (2018). https://docs.docker.com/
15. Dr. Wetter IT-Consulting: testssl.sh (2018). https://github.com/drwet-ter/testssl.sh/
16. Rescorla, E.: The Transport Layer Security (TLS) Protocol Version 1.3, March 2016. https://tools.ietf.org/html/draft-ietf-tls-tls13-12

17. Krawczyk, H., Paterson, K.G., Wee, H.: On the security of the TLS protocol: a systematic analysis. In: CRYPTO 2013, pp. 429–448. Springer (2013)
18. IANA: Transport Layer Security (TLS) Parameters (2018). https://www.iana.org/assignments/tls-parameters/tls-parameters.xhtml. Accessed 10Feb 2018
19. IANA: Transport Layer Security (TLS) Parameters (2005–2015). http://www.iana.org/assignments/tls-parameters
20. IETF: Transport Layer Security (TLS) Extensions, April 2006. http://www.ietf.org/rfc/rfc4366.txt
21. Bhargavan, K., Fournet, C.: Cryptographically verified implementations for TLS. In: Proceedings of the 15th ACM Conference on Computer and Communications Security, CCS 2008, pp. 459–469. ACM (2008)
22. Bhargavan, K., Fournet, C., Kohlweiss, M., Pironti, A., Strub, P.-Y.: Implementing TLS with Verified Cryptographic Security. Proceedings of the 2013 IEEE Symposium on Security and Privacy, SP 2013, May, pp. 445–460. IEEE Computer Society (2013)
23. Hickman, K.E.: The SSL protocol. From Internet Draft, Internet Engineering Task Force, April 1995. http://tools.ietf.org/html/draft-hickman-netscape-ssl-00
24. Rosenfeld, M.: Internet Explorer SSL Vulnerability, May 2018. https://www.thoughtcrime.org/ie-ssl-chain.txt
25. Mozilla: cipherscan (2018). https://github.com/mozilla/cipherscan
26. MozillaWiki: Security/Server Side TLS (2018). https://wiki.mozilla.org/Security/Server_Side_TLS. 10 Febr 2018
27. Mavrogiannopoulos, N., Vercauteren, F.: A cross-protocol attack on the TLS protocol. In: Proceedings of the 2012 ACM conference on Computer and communications security, ser. CCS 2012, pp. 67–72. ACM (2012)
28. AlFardan, N.J., Bernstein, D.J., Paterson, K.G., Poettering, B., Schuldt, J.C.N.: On the security of RC4 in TLS. In: Proceedings of the 22th USENIX Security Symposium, pp. 305–320, July 2013
29. Nginx, January 2018. https://nginx.org/en/
30. OpenSSL Foundation, Inc. (n.d.). https://www.openssl.org/
31. Hoffman, P.: SMTP Service Extension for Secure SMTP over Transport Layer Security. RFC 3207 (Proposed Standard). Updated by RFC 7817, February 2002
32. Bardou, R., Focardi, R., Kawamoto, Y., Simionato, L., Steel, G., Tsay, J.-K.: Efficient Padding Oracle Attacks on Cryptographic Hardware, April 2012. http://hal.inria.fr/hal-00691958
33. RFC 7457: Summarizing Known Attacks On Transport Layer Security (TLS) And Datagram TLS (DTLS), January 2018. https://tools.ietf.org/html/rfc7457
34. Blake-Wilson, S., Nystrom, M., Hopwood, D., Mikkelsen, J., Wright, T.: Transport Layer Security (TLS) Extensions. (Internet Engineering Task Force) From RFC 3546 (Proposed Standard), June 2003. http://www.ietf.org/rfc/rfc3546.txt
35. Turner, S., Polk, T.: Prohibiting Secure Sockets Layer (SSL) Version 2.0. (Internet Engineering Task Force) From RFC 6176 (Proposed Standard), March 2011. http://www.ietf.org/rfc/rfc6176.txt
36. Pornin, T.: TestSSLServer (2017). https://github.com/pornin/TestSSLServer/

Analyses of Characteristics of Changes in Cerebral Activation Status, Depending on Blood Types, in Response to Auditory Stimulation

Jeong Hoon Shin[(✉)] and Hyo Won Jeon

Department of IT Engineering, Daegu Catholic University,
Hayang-Ro 13-13, Hayang-Eup, Gyeongsan-si, Gyeongbuk, Republic of Korea
{only4you, sees9707}@cu.ac.kr

Abstract. As there has recently been growing interest in the correlation between blood type and personality in various fields, including educational organizations, corporations, culture related businesses, etc., numerous studies have been performed, and the results of such studies have been increasingly applied to the aforementioned organizations. In particular, a book authored by Park Hee-Yeon (2006), titled "Blood Type Reveals the Personality", showed that there is a correlation between blood type and temperament, or physical constitutions, which have been proved by results of many investigations and experiments conducted thus far. Based on such facts, it was considered that blood types are likely to affect the brain waves indicative of the status of human cerebral activities, and the purpose of this study was to analyze the patterns of variation of the cerebral activation status, depending on blood types, in response to external stimulation. The results of this study would provide the basis for classification of subjects, which is necessary for the selection of effective auditory stimulant sound sources in the process of neurofeedback therapy/training, inducing the users to modify their own brain waves for healing, and would also present the measures for the selection of efficient visual and auditory stimuli.

Keywords: Blood type · External stimulation · Auditory stimuli

1 Introduction

People in contemporary society are living in a way that suits each individual's ability, social environment, and personal characteristics. Particularly, roles are distributed in a way that is suited for social skills and appropriateness of individuals, as well as individual capabilities, in organizations. Also organizations are formed in light of characteristics and personality of individuals to increase their efficiency. Studies have been conducted on the applications of results thereof to organizations, and particularly, studies investigating the correlation between blood type and personality have been performed.

© Springer Nature Switzerland AG 2019
R. Lee (Ed.): ACIT 2018, SCI 788, pp. 89–103, 2019.
https://doi.org/10.1007/978-3-319-98370-7_8

The results of the study on the relationship between actual human personality and blood type were examined in light of various factors, and the statistical data were presented on the basis of the results of the experiments [1]. However, studies on the relationship between blood type and personality have not yet produced clear conclusions. In addition, there has been little study on the correlation between variations of human brain activities and blood type. Some researchers have claimed that there is a correlation between cerebral activity and blood type, and a correlation between blood type and temperament or physical constitution, and this fact has been confirmed in various investigations and experiments that have been carried out so far [2].

The brain wave is the sum of electrical changes arising from activities of neurons in cerebral cortex, which is recorded with the electrodes attached to the scalp, and represents composite signals that have statistical characteristics according to time and space. [3, 4] These brain wave signals are characterized by the cerebral learning status of individuals, and sensory organs' acceptance of stimuli, etc. The brain waves, analyzed based on such characteristics of the brain, have found wide ranging applications in various fields that encompass the treatment of mental diseases and psychological conditions. Among the brain wave signals that have been an analyzed, it has been found that the brain wave signals of people with mental illness exhibit features different from those of normal people.

In the neurofeedback therapy/training technique, the difference between normal and abnormal brain waves is analyzed, and the users undergo training to change their abnormal brain waves into normal brain waves. This training has enabled control of brain waves previously known as involuntary muscles, and since then, many researchers have made various attempts and experiments on brain waves. [5, 6] However, the subjects have been found to have experienced difficulty in learning how to change their own brain waves. To resolve the problem that subjects cannot easily learn how to change their brain waves, and the inability to apply neurofeedback therapy training techniques as a consequence, some recent research has used neurofeedback/training techniques applying external stimuli such as visual and auditory stimuli. [1] In an improved neurofeedback therapy/training technique, certain auditory and visual stimuli are utilized to induce abnormal brain waves of subjects to change into normal brain waves. However, even if the same external stimuli are applied to the subjects, changes in the cerebral activation state occur in various forms. Such a difference is considered attributable to the difference in characteristics of sensory organs in individual subjects, and differences in their cerebral learning status.

In the improved neurofeedback therapy/training technique, visual and auditory stimuli are required to be applied to induce abnormal brain waves of the subjects change into normal brain waves, and various visual and auditory stimuli may be required depending on the subjects' current condition and characteristics. To apply the optimized visual and auditory stimuli to all subjects at this time, it is necessary to analyze the state of change in the subjects by using a large number of combinations of visual and auditory stimuli, and it takes a long time to perform these tasks, which is considered a disadvantage of neurofeedback therapy/training techniques.

To minimize the time required to derive the visual and auditory stimuli optimized for the subjects, we classified the subject groups into blood types, which is presumed to have an effect on the cerebral activation state, and then exposed candidates to visual and auditory stimuli which was optimized for concerned subjects having been classified into respective groups. Subjects who belonged to a specific candidate group underwent the process where optimal stimuli were selected. At this time, visual and auditory stimuli applied to subjects were classified into the types exclusively applicable to specific candidate groups, and only minimized combinations were utilized. This method helps save significant time in comparison to the process by which visual and auditory stimuli optimized for subjects are selected by subject themselves, and as a result, is expected to find practical application to various fields [7].

In this study, we classified the subjects into candidate subjects based on their blood type and applied the sound of center frequencies, corresponding to the critical band, to each subject [8, 9], and analyzed the changes in brain activation status classified into different categories depending on blood type [10].

2   Experiment Method

In this study, a total of 60 healthy subjects were recruited. Gender, age and possible auditory organ abnormalities were examined in advance. In particular, blood type information was checked during the recruitment of subjects in order to determine the presence of any significant differences between blood type, and variation of cerebral activation status in the analyses. There were 23 subjects with blood type A, 17 with type O, 14 with type B, and 6 with type AB (A > O > B > AB), and subjects were divided into different groups based on their respective blood type. After applying 'various auditory stimuli of previously prepared audible frequency bands' to these subjects, we analyzed the changes in cerebral activation status. Auditory stimulation within the audible frequency band was prepared on the basis of the critical frequency

Table 1. Critical band and center frequencies.

Critical band (center frequency, Hz)	0 ~ 100 (50)	770 ~ 920 (845)	2320 ~ 2700 (2,510)	7700 ~ 9500 (8,600)
	100 ~ 200 (150)	920 ~ 1080 (1,000)	2700 ~ 3150 (2,925)	9500 ~ 12000 (10,750)
	200 ~ 300 (250)	1080 ~ 1270 (1,175)	3150 ~ 3700 (3,425)	12000 ~ 15500 (13,750)
	300 ~ 400 (350)	1270 ~ 1480 (1,375)	3700 ~ 4400 (4,050)	15500 ~ 22500 (19,000)
	400 ~ 510 (455)	1480 ~ 1720 (1,600)	4400 ~ 5300 (4,850)	
	510 ~ 630 (570)	1720 ~ 2000 (1,860)	5300 ~ 6400 (5,850)	
	630 ~ 770 (700)	2000 ~ 2320 (2,160)	6400 ~ 7700 (7,050)	

band in which the auditory organ responded sensitively in order to maximize the variations of cerebral activation status. In this study, we utilized the center frequency of 25 critical bands. Table 1 below presented the center frequency of each critical band selected in this study.

Using the auditory stimulant sound sources presented in Table 1, we gathered brain wave signals for analysis of changes in cerebral activation status after applying auditory stimuli to subjects. At this time, the experiment was conducted in an independent experimental space to minimize miscellaneous wave generation caused by external influences. A total of 10 points, including F7, F8, T3, T4, T5, T6, Fz and Cz, and reference electrodes A1 and A2, were arranged in accordance with the International Electrodes System when the electrodes for brain wave collection were placed. When analog signals were converted to digital signals, a sampling rate of 256 Hz was applied, in order to minimize data loss. Gathering and analyses of brain wave signals from the subjects were performed through the following procedures:

1. Electrodes were attached to the preselected positions in an independent experimental environment (F7, F8, T3, T4, T5, T6, CZ, FZ, A1, A2).
2. After collecting the brain wave signals of a stable section from the subjects, the brain wave signals that reflect the cerebral activation status based on auditory stimulus were collected.
3. When 25 types of stimulant sound sources were applied, the brain waves were collected for 10 s per sound source, in order to analyze the status of response to stimulation.
4. The collected brain wave signals were analyzed quantitatively by channel (the mean relative energy analyses were performed on each of 8 channels, except for reference electrodes).
5. In the quantitative analysis, the reference band of brain wave signals used in this study were based on 6 frequency bands including delta, theta, alpha, beta, gamma, and SMR waves.
6. Using the results of the mean relative energy analysis, the presence of specific sound sources and characteristics and reactions of subjects were analyzed.
7. After repeating the processes from 1 to 6 for 60 subjects, the results of the statistical analyses were calculated.

3   Results of Experiment

In this study, we carried out an experiment to analyze the correlation between blood type and brain waves showing the cerebral activation status. For that, brain wave signals were collected at the same time, while the prepared stimulant sound sources of 25 critical bands were applied to each subject. The cerebral activation status of the 6 bands (delta, theta, alpha, beta, gamma, and SMR bands) was analyzed by using the collected signals. At this time, the critical band frequencies of the auditory stimuli that generated the highest relative energy and the lowest relative energy for each band were identified from the subjects, and then common characteristics and differences by blood type were analyzed. For example, as shown in Fig. 1, brain wave signals were generated in response to auditory stimuli when stimulant sound sources of 25 critical band frequencies were applied consecutively to the subject A. Based on the delta wave band of CH 1, it could be found that the stimulant sound source of 1000 Hz band generated the 'maximum mean relative energy' of delta waves in the subject. In addition, the stimulant sound source of 455 Hz band was found to generate the 'minimum mean relative energy' of delta waves in the subject.

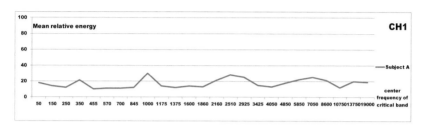

Fig. 1. Graph of mean relative energy of delta waves by stimulant sound source of critical band in subject A (CH 1).

In this case, we defined the stimulant sound source in the 1000 Hz band as the delta maximum, and the stimulant sound source in the 455 Hz band as the delta minimum for subject A in this study.

Figure 2 is a graphical representation of the maximum/minimum (Max/Min) distribution by stimulant sound source for analyses of statistical reproducibility. For example, in the upper left graph (CH 1) of Fig. 2-a, 2 subjects with blood type A, 1 subject with blood type, and zero subjects with blood type O and blood type AB were found to respond to stimulant sound source of the 570 Hz frequency band corresponding to the delta maximum. In this study, the statistical reproducibility was considered to be valid when the number of subjects who exhibited Max/Min values for specific sound sources comprised at least 15% (20% or higher for subjects with blood type AB) of the population by blood type.

According to Fig. 2-a, the statistical reproducibility of the delta Max is found in all blood type groups in response to the stimulant sound source in the 50 Hz frequency band. Particularly, in the subjects with blood type B, the statistical reproducibility of the delta Max was found to exist in CH 3, CH 4, and CH 8 by stimulant sound sources of the 150 Hz frequency band, in CH 8 by the stimulant sound source of the 250 Hz frequency band, and in CH 6 by stimulant sound sources of the 2510 Hz frequency band.

Meanwhile, the statistical reproducibility of the delta Max was found to exist in subjects with blood type O in CH 2 by stimulant sound sources of the 4050 Hz frequency band, as well as 50 Hz. Furthermore, the statistical reproducibility of the delta Max was found to exist in CH 2, CH 3, CH 4, CH 6, and CH 8 by stimulant sound sources of the 13750 Hz frequency band. In the Fig. 2-b, the statistical reproducibility of the delta Max was found to exist in CH 3 by stimulant sound sources of the 1375 Hz frequency band and in CH 6 by stimulant sound sources of the 1600 Hz frequency band among subjects with blood type A.

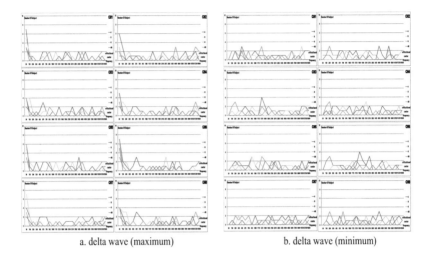

a. delta wave (maximum)   b. delta wave (minimum)

Fig. 2. Statistical reproducibility for delta wave (blood type A (male and female): 23 persons, blood type B (male and female): 14 persons, blood type O (male and female): 17 persons, blood type AB (male and female): 6 persons).

In the subjects with blood type B, the statistical reproducibility of the delta Max was found to exist also in CH 3 by stimulant sound sources of the 250 Hz frequency band and in CH 5 by stimulant sound sources of the 845 Hz frequency band among the subjects with blood type A. In addition, the statistical reproducibility of the delta Max was found to exist in CH 4 by stimulant sound sources of the 2160 Hz frequency band and CH 1 by stimulant sound sources of the 10750 Hz frequency band in this subject group.

In the subjects with blood type O, the statistical reproducibility of the delta Min was found to exist in 2, 3, 4, 6, and 8 CH by stimulant sound sources of the 250 Hz frequency band and also in CH 3 by stimulant sound sources of the 750 Hz frequency band among the subjects with blood type A. Additionally, the statistical reproducibility of the delta Max was found to exist in CH 1 and CH 5 by stimulant sound sources of the 1375 Hz frequency band in this subject group.

In the subjects with blood type AB, the statistical reproducibility of the delta Max was found to exist also in CH 1 by stimulant sound sources of the 455 Hz frequency band and in CH 5 and CH 7 by stimulant sound sources of the 2160 Hz frequency band. Moreover, the statistical reproducibility of the delta Max was found to exist also in CH 8 by stimulant sound sources of the 8600 Hz frequency band in this subject group.

Figure 3 is a graphical representation of the Max/Min distribution by stimulant sound sources for the analyses of statistical reproducibility for theta waves. As shown in the Fig. 3-a, statistical reproducibility of theta Max was found to exist in CH 1 and CH 7 by stimulant sound sources of the 50 Hz frequency band, and the statistical reproducibility was found to exist in CH 2 and CH 7 by stimulant sound sources of the 455 Hz frequency band among subjects with blood type A. Besides, the statistical reproducibility of theta Max was found to exist in CH 8 by stimulant sound sources of the 10750 Hz frequency band in this subject group.

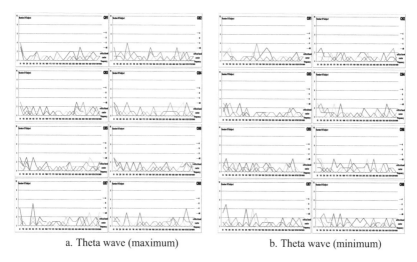

a. Theta wave (maximum)　　　　b. Theta wave (minimum)

Fig. 3. Statistical reproducibility for theta wave (blood type A (male and female): 23 persons, blood type B (male and female): 14 persons, blood type O (male and female): 17 persons, blood type AB (male and female): 6 persons).

In the subjects with blood type B, statistical reproducibility of theta Max was found to exist in [CH 1 and CH 2], and [CH 4, CH 5 and CH 7] by stimulant sound sources of the 50 Hz frequency band, and by stimulant sound sources of the 8600 Hz frequency band, respectively. In addition, statistical reproducibility of theta Max was found to exist in CH 7 by stimulant sound sources of the 19000 Hz frequency band in this subject group.

In the subjects with blood type O, statistical reproducibility of theta Max was found to exist in CH 1, CH 3, and CH 6 by stimulant sound sources of the 50 Hz frequency band and in CH 3 by stimulant sound sources of the 5850 Hz frequency band. Additionally, statistical reproducibility of theta Max was found to exist in CH 4 by stimulant sound sources of the 4050 Hz frequency band in this subject group.

In the subjects with blood type AB, the statistical reproducibility of the delta Max was found to exist in all channels, excluding CH 5 and CH 6, by stimulant sound sources of the 50 Hz frequency band and in CH 5 by stimulant sound sources of the 250 Hz frequency band, and furthermore, in CH 6 by stimulant sound sources of the 19000 Hz frequency band.

As shown in Fig. 3-b, statistical reproducibility of theta Min was found to exist in CH 7 by stimulant sound sources of the 150 Hz frequency band and also in CH 7 by stimulant sound sources of the 845 Hz frequency band among subjects with blood type A.

In the subjects with blood type B, statistical reproducibility of theta Min was found to exist in CH 4 and CH 6 by stimulant sound sources of the 50 Hz frequency band and in CH 6 by stimulant sound sources of the 150 Hz frequency band, and furthermore, in CH 1 by stimulant sound sources of the 250 Hz frequency band. In addition, statistical reproducibility of theta Min was found to exist in CH 4, CH 7, and CH 6 by stimulant sound sources of the 570 Hz, 1175 Hz, and 2925 Hz frequency bands, respectively, in this subject group.

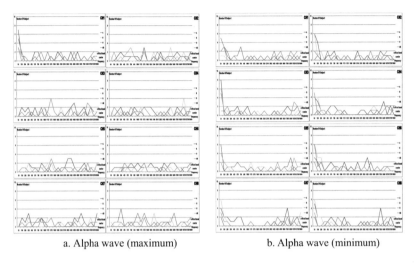

a. Alpha wave (maximum)   b. Alpha wave (minimum)

Fig. 4. Statistical reproducibility for alpha wave (blood type A (male and female): 23 persons, blood type B (male and female): 14 persons, blood type O (male and female): 17 persons, blood type AB (male and female): 6 persons).

In the subjects with blood type O, statistical reproducibility of theta Min was found to exist in CH 3 and CH 6 by stimulant sound sources of the 50 Hz frequency band, in CH 3 and CH 8 by stimulant sound sources of the 250 Hz frequency band, in CH 4 by stimulant sound sources of the 570 Hz frequency band, and in CH 1, CH 6, CH, and 7 CH of the 1375 Hz frequency band. In addition, statistical reproducibility of theta Min was found to exist in CH 2 by stimulant sound sources of the 13750 Hz frequency band in this subject group.

In the subjects with blood type AB, statistical reproducibility of theta Min was found to exist in CH 3 by stimulant sound sources of the 50 Hz frequency band and in CH 2 by stimulant sound sources of the 1000 Hz frequency band. Besides, statistical reproducibility of theta Min was found to exist in CH 8 by stimulant sound sources of the 1175 Hz frequency band and in CH 6 by stimulant sound sources of the 4050 Hz frequency band in this subject group.

Figure 4 is a graphical representation of the Max/Min distribution used to perform analyses of statistical reproducibility for alpha waves. As shown in Fig. 4-a, statistical reproducibility of alpha Max was found to exist in CH 1 by stimulant sound sources of the 50 Hz frequency band and in CH 8 by stimulant sound sources of the 350 Hz frequency band.

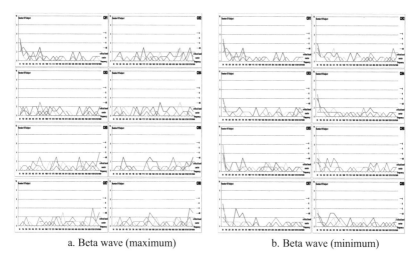

a. Beta wave (maximum)   b. Beta wave (minimum)

Fig. 5. Statistical reproducibility for beta wave (blood type A (male and female): 23 persons, blood type B (male and female): 14 persons, blood type O (male and female): 17 persons, blood type AB (male and female): 6 persons).

In the subjects with blood type B, statistical reproducibility of alpha Max was found to exist in CH 1 by stimulant sound sources of the 50 Hz frequency band and in CH 3 by stimulant sound sources of the 2160 Hz frequency band. Moreover, statistical reproducibility of alpha Max was found to exist in CH 2 and CH 4 by stimulant sound sources of the 4850 Hz frequency band and stimulant sound sources of the 10750 Hz frequency band, respectively, in this subject group.

In the subjects with blood type O, statistical reproducibility of alpha Max was found to exist in CH 1 by stimulant sound sources of the 50 Hz frequency band and in CH 2 by stimulant sound sources of the 700 Hz frequency band, and in CH 3 and CH by stimulant sound sources of the 1375 Hz frequency band, and furthermore, in CH 3 by stimulant sound sources of the 8600 Hz frequency band.

As shown in Fig. 4-b, statistical reproducibility of alpha Min is found to exist by stimulant sound sources of the 50 Hz frequency band in all blood types. In the subjects with blood type A, statistical reproducibility of alpha Min was found to exist in CH 2 by stimulant sound sources of the 150 Hz frequency band and in CH 7 and CH 8 by stimulant sound sources of the 7050 Hz frequency band.

In the subjects with blood type B, statistical reproducibility of alpha Min was found to exist in CH 1, CH 4, and CH 6 by stimulant sound sources of the 150 Hz frequency band. In addition, statistical reproducibility of alpha Min was found to exist in CH 6 by stimulant sound sources of the 250 Hz frequency band and in CH 7 by stimulant sound sources of the 19000 Hz frequency band in this subject group.

In the subjects with blood type O, statistical reproducibility of alpha Min was found to exist in CH 1 by stimulant sound sources of the 150 Hz frequency band and in CH 3 and CH 5 by stimulant sound sources of the 10750 Hz frequency band.

In the subjects with blood type AB, statistical reproducibility of alpha Min was found to exist in CH 2 by stimulant sound sources of the 455 Hz frequency band and in CH 4 by stimulant sound sources of the 19000 Hz frequency band.

Figure 5 is a graphical representation of the Max/Min distribution by stimulant sound sources, used to perform analyses of statistical reproducibility for beta waves. As shown in Fig. 5-a, statistical reproducibility of beta Max was found to exist in CH 1 by stimulant sound sources of the 50 Hz frequency band in the subjects with blood type A, and furthermore, the statistical reproducibility of beta Max was also found to exist in CH 8 by stimulant sound sources of the 13750 Hz frequency band and in CH 5 by stimulant sound sources of the 19000 Hz frequency band.

In the subjects with blood type B, statistical reproducibility of beta Max was found to exist in CH 1 by stimulant sound sources of the 150 Hz frequency band, and in CH 7 by stimulant sound sources of the 2160 Hz frequency band, and furthermore, in CH 2 and CH 4 by stimulant sound sources of the 7050 Hz frequency band.

In the subjects with blood type O, statistical reproducibility of beta Max was found to exist in CH 1 by stimulant sound sources of the 50 Hz frequency band. In addition, statistical reproducibility of beta Max was found to exist in CH 3 and CH 5 by stimulant sound sources of the 700 Hz frequency band, in CH 6 by stimulant sound sources of the 4850 Hz frequency band, and in CH 7 by stimulant sound sources of the 10750 Hz frequency band in this subject group.

As shown in Fig. 5-b, statistical reproducibility of beta Min was found to exist in CH 1, CH 2, CH 3, CH 4, CH 5, and CH 7 by stimulant sound sources of the 50 Hz frequency band and in CH 7 by stimulant sound sources of the 455 Hz frequency band in the subjects with blood type A.

In the subjects with blood type B, statistical reproducibility of beta Min was found to exist in CH 3, CH 6, CH 7, and CH 8 by stimulant sound sources of the 50 Hz frequency band. Additionally, statistical reproducibility of beta Min was found to exist in CH 1 by stimulant sound sources of the 150 Hz frequency band and in CH2 by stimulant sound sources of the 570 Hz frequency band in the subject group.

In subjects with blood type O, statistical reproducibility of beta Min was found to exist in all channels, excluding CH 3 and CH 7, by stimulant sound sources of the 50 Hz frequency band and stimulant sound sources of the 13750 Hz frequency band, respectively.

In the subjects with blood type AB, statistical reproducibility of beta Min was found to exist in CH 1, CH 2, CH 5, and CH 6 by stimulant sound sources of the 50 Hz frequency band.

Figure 6 is a graphical representation of the Max/Min distribution by stimulant sound source, used to perform analyses of statistical reproducibility for gamma waves. As shown in Fig. 6-a, statistical reproducibility of gamma Max was found to exist in CH 7 by stimulant sound sources of the 50 Hz frequency band, in CH 5 by stimulant sound sources of the 150 Hz frequency band, and CH 7 by stimulant sound sources of the 455 Hz frequency band in subjects with blood type A.

In subjects with blood type B, statistical reproducibility of gamma Max was found to exist in CH3, CH 4, CH 5, CH 6, CH 7, and CH 8 by stimulant sound sources of the 50 Hz frequency band, and furthermore, statistical reproducibility of gamma Max was found to exist in CH 6 by stimulant sound sources of the 1175 Hz frequency band, and

CH 4 by stimulant sound sources of the 4850 Hz frequency band. Additionally, in subjects with blood type B, statistical reproducibility of gamma Max was also found to exist in CH 1 by stimulant sound sources of the 10750 Hz frequency band, and in CH 1 and CH 2 by stimulant sound sources of the 19000 Hz frequency band.

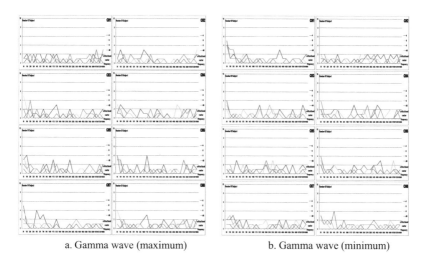

a. Gamma wave (maximum)   b. Gamma wave (minimum)

Fig. 6. Statistical reproducibility for gamma wave (blood type A (male and female): 23 persons, blood type B (male and female): 14 persons, blood type O (male and female): 17 persons, blood type AB (male and female): 6 persons).

In subjects with blood type O, statistical reproducibility of gamma Max was found to exist in CH 6 by stimulant sound sources of the 50 Hz frequency band, in CH 3 by stimulant sound sources of the 250 Hz frequency band, and in CH 6 by stimulant sound sources of the 7050 Hz frequency band.

In subjects with blood type AB, statistical reproducibility of gamma Max was found to exist in CH 3 by stimulant sound sources of the 250 Hz frequency band, in CH 5 by stimulant sound sources of the 700 Hz frequency band, and in CH 4 by stimulant sound sources of the 10750 Hz frequency band.

As shown in the Fig. 6-b, in subjects with blood type A, statistical reproducibility of gamma Min was found to exist in CH 1 and CH 3 by stimulant sound sources of the 50 Hz frequency band, and furthermore, statistical reproducibility of gamma Min was found to exist in CH 8 by stimulant sound sources of the 455 Hz frequency band.

In the subjects with blood type B, statistical reproducibility of gamma Min was found to exist in CH 1 and CH 6 by stimulant sound sources of the 50 Hz frequency band, and also in CH 1 by stimulant sound sources of the 150 Hz and 250 Hz frequency bands, respectively. Moreover, statistical reproducibility of gamma Min was also found to exist in CH 5 by stimulant sound sources of the 860 Hz frequency band, and in CH 6 by stimulant sound sources of the 10750 Hz frequency band.

In subjects with blood type O, statistical reproducibility of gamma Min was found to exist in CH 1, CH 3, CH 4, and CH 6 by stimulant sound sources of the 50 Hz frequency band, and in CH 7 and CH 5 by stimulant sound sources of the 150 Hz and 1860 Hz frequency bands, respectively. In addition, statistical reproducibility of gamma Min was found to exist in CH 3 by stimulant sound sources of the 2160 Hz frequency band, and in CH 4 by stimulant sound sources of the 2510 Hz frequency band in this subject group.

In subjects with blood type AB, statistical reproducibility of gamma Min was found to exist in CH 6, CH 1, CH 8, and CH 5 by stimulant sound sources of 50 Hz, 570 Hz, 845 Hz, and 2925 Hz frequency bands, respectively. Additionally, reproducibility of gamma Min was found to exist in CH 4 and CH 6 by stimulant sound sources of the 3425 Hz frequency band.

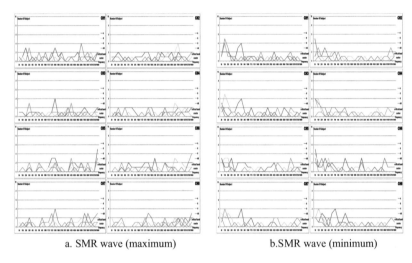

a. SMR wave (maximum)   b.SMR wave (minimum)

Fig. 7. Statistical reproducibility for SMR wave (blood type A (male and female): 23 persons, blood type B (male and female): 14 persons, blood type O (male and female): 17 persons, blood type AB (male and female): 6 persons).

Figure 7 is a graphical representation of the Max/Min distribution by stimulant sound sources, used to perform analyses of statistical reproducibility for SMR waves. As shown in the Fig. 7-a, in the subjects with blood type A, statistical reproducibility of SMR Max was found to exist in CH 3 by stimulant sound sources of the 1600 Hz frequency band, and in CH 5 by stimulant sound sources of the 19000 Hz frequency band.

In the subjects with blood type B, statistical reproducibility of SMR Max was found to exist in CH 7 by stimulant sound sources of the 1600 Hz frequency band, in CH 1, CH 4, and CH 6 by stimulant sound sources of the 5850 Hz frequency band, and in CH 2 by stimulant sound sources of the 7050 Hz frequency band. In subjects with blood type O, statistical reproducibility of SMR Max was found to exist in CH 1 and CH 3 by stimulant sound sources of the 350 Hz frequency band, and furthermore, statistical reproducibility of SMR Max was found to exist in CH 7 by stimulant sound sources of the 1375 Hz frequency band, and in CH 3 and CH 5 by stimulant sound sources of the 3425 Hz frequency band. Moreover, statistical reproducibility of SMR Max was found to exist in CH 5 and CH 6 by stimulant sound sources of the 19000 Hz frequency band.

In the subjects with blood type AB, statistical reproducibility of SMR Max was found to exist in CH 3 and CH 5 by stimulant sound sources of the 2925 Hz and 5850 Hz frequency bands, respectively. Besides, statistical reproducibility of SMR Max was found to exist in CH 7 and CH 8 by stimulant sound sources of the 8600 Hz frequency band, and in CH 1 by stimulant sound sources of the 10750 Hz frequency band.

Table 2. Average distribution of mean relative energy of each waveform by blood type

	Max/Min Hz(CH)	blood type A	blood type B	blood type O	blood type AB
DELTA	Max	50(all)	50(all),150(3,4,8),250(8),2510(6)	50(1,3,4,5,6,7,8),4050(2),13750(2,4,8)	50(1,3,4,5,6,7,8)
	Min	1375(3),1600(6)	250(3),845(5),2160(4),10750(1)	250(2,3,4,6,8),750(8),1375(1,5)	455(1),2160(5,7),8600(8)
THETA	Max	50(1,7),455(2,7),10750(8)	50(1,2),8600(4,5,7),19000(7)	50(1,3,6),5850(3),4050(4)	50(1,2,3,4,7,8),250(5),19000(6)
	Min	150(7),845(7)	50(4,6),150(6),250(1),570(4),1175(7),2925(6)	50(3,6),250(3,8),570(4),1375(1,6,7),13750(2)	50(3),1000(2),1175(8),4050(6)
ALPHA	Max	50(1),350(8)	50(1),2160(3),4850(2),10750(4)	50(1),700(2),1375(3,5),8600(3)	50(1),455(6),845(3,7)
	Min	50(1,2,3,4,5,7,8),150(2),7050(7,8)	50(1,3,4,5,6,7,8),150(1,4,6),250(6),19000(7)	50(3,4,5,6,7,8),150(1),10750(3,5)	50(5,6,7,8),455(2),19000(4)
BETA	Max	50(1),13750(8),19000(5)	150(1),2160(7),7050(2,4)	50(1),700(3,5),4850(6),10750(7)	50(1),150(3),8600(2)
	Min	50(1,2,3,4,5,7),455(7)	50(3,6,7,8),150(1),570(2)	50(1,2,4,5,6,8),13750(2)	50(1,2,5,6)
GAMMA	Max	50(7),150(5),455(7)	50(3,4,5,6,7,8),1175(6),4850(4),10750(1),19000(1,2)	50(6),250(3),7050(6)	250(3),700(5),10750(4)
	Min	50(1,3),455(8)	50(1,6),150(1),250(1),1860(5),10750(6)	50(1,3,4,6),150(7),1860(5),2160(3),2510(4)	50(6),570(1),845(8),2925(5),3425(4,6)
SMR	Max	1600(3),19000(5)	1600(7),5850(1,4,6),7050(2)	350(1,3),1375(7),3425(3,5),19000(5,6)	2925(3),5850(5),8600(7,8),10750(1)
	Min	50(2,4,6),700(1,3,7,8)	50(6,7),150(1,3,4,5,8),250(1,3,5,7),455(6)	50(4),150(1,2,7),250(6),8600(5,6),13750(2)	50(3,5,6,7,8)

As shown in the Fig. 7-b, in subjects with blood type A, statistical reproducibility of SMR Min was found to exist in CH 2, CH 4, and CH 6 by stimulant sound sources of the 50 Hz frequency band, and in CH 1, CH 3, CH 7, and CH 8 by stimulant sound sources of the 700 Hz frequency band.

In subjects with blood type B, statistical reproducibility of SMR Min was found to exist in CH 6 and CH 7 by stimulant sound sources of the 50 Hz frequency band, in CH 1, CH 3, CH 4, CH 5, and CH8 by stimulant sound sources of the 150 Hz frequency band, in CH 1, CH 3, CH 5, and CH 7 by stimulant sound sources of the 250 Hz frequency band, and in CH 6 by stimulant sound sources of the 455 Hz frequency band.

In subjects with blood type O, statistical reproducibility of SMR Min was found to exist in CH 4 by stimulant sound sources of the 50 Hz frequency band, and in CH 1, CH 2, and CH 7 by stimulant sound sources of the 150 Hz frequency band. Additionally, statistical reproducibility of SMR Min was found to exist in CH 2 by stimulant sound sources of the 13750 Hz frequency band, as well as in CH 6 by stimulant sound sources of the 250 Hz frequency band, and in CH 5 and CH 6 by stimulant sound sources of the 8600 Hz frequency band.

In the subjects with blood type AB, statistical reproducibility of SMR Min was found to exist in CH 3, CH 5, CH 6, CH 7, and CH 8 by stimulant sound sources of the 50 Hz frequency band.

4   Conclusion

In this study, we analyzed the changes in cerebral activation status, depending on blood type and auditory stimuli, and carried out experiments to determine common and different characteristics based on blood type, the results of which could be applied to neurofeedback therapy/training.

In relation to the changes in brain wave signals of subjects who belonged to the same blood type group, the results of the experiments in this study suggested that the auditory stimulant sound sources applied to the subjects caused activation or non-activation of specific frequency bands. If these results are applied well, a door will be opened to neurofeedback therapy/training for the users who experience difficulty in making themselves familiarized with neurofeedback therapy/training.

In this study, external stimulant sounds were used which could minimize the difference in brain waves between subjects and normal people, to resolve such difficulties as described above, so that the difference between brain waves of subjects, and reference brain waves of normal people could be minimized without the need for active involvement of subjects.

For example, if there were any subjects who needed to reduce the delta waves in the temporal lobe of the left hemisphere of the brain, in accordance with the protocol of neurofeedback therapy, the relative energy value of the delta wave band could be reduced in the temporal lobe of the left hemisphere of the brain by applying stimulant sound sources of the 1375 Hz band to subjects with blood type A, and by applying stimulant sound sources of the 250 Hz band to subjects with blood type B, and subjects with blood type O, as shown in Table 2.

Acknowledgments. This work was supported by the sabbatical research grant from Daegu Catholic University in 2015.

References

1. Shin, J.H., Jeon, H.Y.: Analyses of characteristics of the changes in cerebral activation state based on sasang-constitution. Int. Inf. Inst. **20**(9), 6645–6656 (2017)
2. Hee-Heyon, P.: Blood types provide window into the personality of person, East and West, South Korea (2006)
3. Byeok-Jin, L., Seon-Guk, Y.: Analyses of correlation between space and frequency in brain-induced potentials based on concentration of visual stimulation. J. Korea Instit. Electron. Inf. Eng. **50**(10), 2017–2028 (2013)
4. Broughton, R., Hasan, J.: Quantitative topographic electroencephalographic mapping during drowsiness and sleep onset. J. Clin. Neurophysiol. Off. Publ. Am. Electroencephalogr. Soc. **12**(4), 372–386 (1995)
5. Towle, V.L., Bolanos, J., Suarez, D., Tan, K., Grzeszczuk, R., Levin, D.N.: The spatial location of BRAIN WAVE electrodes: locating the best-fitting sphere relative to cortical anatomy. Electroencephalogr. Clin. Neurophysiol. **86**(1), 1–6 (1993)
6. Roy, V., Shukla, S.: A survey on artifacts detection techniques for electro-encephalography (BRAIN WAVE) signals. IJMUE **10**(3), 425–442 (2015)
7. Shin, J., Kim, M., Lee, S.: Characteristic analysis of auditory stimuli utilizing the sound source of ultrasonic band and the audible frequency band and cerebral activation state changes. Int. J. Multimedia Ubiquitous Eng. **10**(10), 315–328 (2015)
8. Shin, J., Kim, M., Lee, S.: Characteristic analysis of auditory stimulation. In: Ubiquitous Science and Engineering 2015, Jeju, Korea, 19–21 August 2015
9. Hosoi, H., Imaizumi, S., Sakaguchi, T., Tonoike, M., Murata, K.: Activation of the auditory cortex by ultrasound. Lancet **351**(9101), 496–497 (1998)
10. Chung-Sik, K., Seon-Gyu, L.: A study on correlation among blood type, personality and stress-resistance in adults. J. Korea Academia-Ind. Coop. Soc. **6**(12), 2554–2560 (2011)

An International Comparative Study on the Intension to Using Crypto-Currency

Kyung-Jin Jung, Jung-Boem Park, Nhu Quynh Phan,
Chen Bo, and Gwang-yong Gim[✉]

Business Administration, Soongsil University, Seoul, Korea
seealove@naver.com, mecs90@gmail.com,
Phanhuquynh@hotmail.com, 1146610225@qq.com,
gygim@ssu.ac.kr

Abstract. Currently, various cryptocurrencies are being traded throughout the day for the whole 24 h via crypto-currency markets in this borderless virtual world. Many people invest in cryptocurrencies which can be utilized and replace existing currencies. Despite this, not many studies on actual demand for cryptocurrencies have been conducted yet. This research extensively applied UTAUT to study elements influencing intention of usage for cryptocurrencies. It also added economic feasibility, payment convenience, government regulation, and risk as independent variables and established an UTAUT model applying said independent variables. Particularly, this research seeks to figure how characteristics among countries respond to the intention of usage for cryptocurrencies by comparing Korea, China, and Vietnam.

Keywords: Crypto-currency · UTAUT · Korea · China · Vietnam

1 Introduction

Currently, various cryptocurrencies are being throughout the day for the whole 24 h via cryptocurrency markets in this borderless virtual world. Cryptocurrency market is completely digitalized differing from existing currency trading markets and crypto-assets including cryptocurrencies are being traded in the market.

Various cryptos with 11,389 types are being traded at exchanges in the world. Major cryptocurrencies include Bitcoin, Ethereum, Ripple, Litecoin, etc. Among these, Bitcoin is presumed to be the financial asset being traded most actively. Furthermore, preceding researches are concerned of cryptocurrencies being used as a haven for risks including hedge, safe asset, capital control, etc. also concerned with cryptocurrencies causing capital market distortion and foreign market disruption.

Price of cryptocurrencies and stocks have rejected simultaneously, which highlighted the issue with stability of cryptocurrencies. Moreover, there is a rising voice saying that cryptocurrencies are not the financial innovative paradigm, but a means of temporary speculation and gambling investment with crypto investors aiming to make marginal profits. Futhermore, there is a rising voice asking for regulating cryptocurrencies for their instability and harmful effects.

© Springer Nature Switzerland AG 2019
R. Lee (Ed.): ACIT 2018, SCI 788, pp. 104–123, 2019.
https://doi.org/10.1007/978-3-319-98370-7_9

Despite rapidly increasing number of malicious acts such as pyramid sales and hackings for cryptocurrency traders inside and outside of the country (Korea, China, Vietnam), no regulations for cryptocurrency trading are set and exist under the current law.

There are two tight stances for crypto markets. First stance asks for adequate regulation for the abnormally hyped trading markets (Korea, China, Vietnam) that are accelerating negative phenomenon. Another stance is that early adoption of regulation may constrict the new 4th industrial revolution market as definitions and concepts for cryptocurrencies are unclear and wide.

It is certain that cryptocurrency markets will remain as a significant phenomenon. The problem is how we perceive such phenomenon.

In terms of social-technological phenomenon, if a technological phenomenon were to be absorbed into the social system, it needs to satisfy all complex social-technological elements such as economic feasibility, payment convenience, government regulation, and risks.

In comparison to interests and policies of each country (Korea, China, Vietnam), the current status is that it still lacks research studying the popular intention of accepting cryptocurrencies and comparison among countries.

To fill this insufficient gap, we believe that UTAUT, a model that is believed to best explain the current acceptance model, is required and should be utilized for conducting actual research for acceptance.

This research studies cryptocurrencies such as Bitcoin. Additionally, various elements affecting the actual acceptance will be figured and this study seeks to understand what routes these cryptocurrencies use to reach to the acceptance. Moreover, through multigroup effect analysis, we seek to understand what differences there are among factors that affect acceptance intention for cryptocurrencies by each country (Korea, China, Vietnam). Thus, based on these ideas, this paper aims to provide both scholastic and practical views for effective utility methods of cryptocurrency trading and usage and directions that technologies should head toward.

To achieve the said goals, in this research, we organized the recent status of cryptocurrencies. Additionally, we looked at theories and models for preceding technical acceptance as we aim to develop an acceptance model based on UTAUT. For this, we provide hypothesis as we review factors preceded for independent variables of existing UTAUT.

2   Literature Review

2.1   Crypto-Currency

Cryptocurrency, is a type of crypto-based currency, and is different from traditional legal tenders, in terms of its issuing party, issuing country, and physical existence. Speaking of legal tenders, centralized issuing country exists, there clearly exists the issuing country, and they are legally assigned physical and substantial currencies, composed of either bills or coins.

For peer-to-peer or P2P online transaction without centralized server, a digital currency called a token is transferred as it delivers all other participants a transaction message saying that an individual is transferring to another party. Other participants who received this transaction message automatically update as they fit the contents of transaction record ledgers they own into this message. In the previous P2P network, it was difficult to prevent double payment in which a same token is used for two times. However, the blockchain technology is the first complete P2P transaction system that resolved the issue with double payment without any managing parties such as the central server, etc. Since everyone has the right to access for transaction ledgers, it belongs to a public blockchain.

Blockchain-based new financial assets are being traded at institutional financial markets such as Chicago Board Option Exchange and other exchanges (CBOE, CME, etc.) In December 2017, Bitcoin Future Trading and derivative products using Bitcoin as assets were released. According to Pieters [10], among cryptocurrencies, Bitcoin is the financial asset being traded the most and is the sole medium of exchange. Furthermore, the total market cap of Bitcoin at December 2017, today, is at 764 billion dollars, or 55% of the whole cryptocurrency trading.

Speaking of advantages that a cryptocurrency has, it not only fulfills its role as the economical 'currency', but it also is useful for protecting privacy and activating electric commercial transactions. Moreover, it reduces transaction costs as well as production costs for issuing currencies. However, due to its anonymity, it can be maliciously used like avoiding taxes. Another shortcoming is that its exchange rate is unstable (IBK Economy Research Center, 2013).

In the case of South Korea, it executed real-name cryptocurrency trading system to strictly respond to financial crimes such as asset laundering and tax avoidance. China banned Initial Coin Offering (ICO) on September 2017 and shut down cryptocurrency exchanges. Moreover, on January, 2018, it banned people from accessing to international and domestic exchanges and platforms that offer cryptocurrency trading services.

State Bank of Vietnam stated that cryptocurrencies such as Bitcoin are not legal methods under regulation and will coordinate with the police department to stop the usage of cryptocurrencies, adding that it will punish those who use and distribute cryptocurrencies based on the criminal law.

2.2 UTAUT (Unified Theory of Acceptance and Use of Technology) Model

Out of research methods for user's intention of usage and continuous use following the new emergence of innovative products and services equipped with IT technologies, TAM (Technology Acceptance Model) is the most basic research model that has been used for the longest term. The model started to expand as technological environment changes and social psychology were reflected into it as time passes by [20].

UTAUT (Unified Theory of Acceptance and Use of Technology) shown in the Fig. 1 was developed by summing 8 types of models regarding the intention of use in order to complement the limits that existing TAMS could not sufficiently reflect many variables. As a result, its explanation ability has increased approximately by 20 to 30% when compared to previous acceptance models [2].

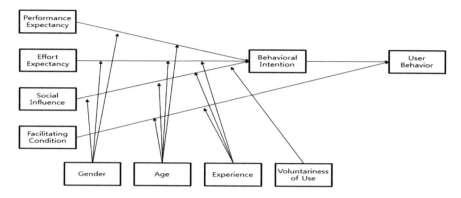

Fig. 1. TAUT model (Venkatesh et al., 2003)

For independent variables of UTAUT model, there is performance expectancy similar to perceived usefulness, effort expectancy, an identical concept to perceived usefulness, social influence and facilitating conditions. Behavioral intention is composed as a parameter and use behavior is used as a dependent variable. Performance expectancy is the degree of usefulness that a user expects performance to be enhanced as he or she uses a system. Effort expectancy is the degree of usefulness for using a system that a user feels easy when using the system. Social influence refers to when a user recommends social atmosphere or neighboring people to use the system and it can also refer to the degree of belief that justifies such usage. Facilitating condition directly posing influence on behavioral use refers to organizational or technological factors that support users when using a new system. Moreover, moderating variables are believed to serve a role when sex, age, experience, and spontaneity of usage influence performance expectancy and effort expectancy and when social influence and faciliating condition influence to the intention of behavior and behavioral usage [20].

This research studies usage intention for cryptocurrencies for Korea, China, and Vietnam. Thus, it conducted a practical research by developing a research model based on UTAUT, a model that is believed to be appropriate for researching external social and faciliating factors as well as technological factors of cryptocurrencies.

2.3   Perceived Risk Theory

The theory of perceived risk first introduced by Bauer [30] refers to the belief that an individual may not know what results would be caused from uncertainties for individual's certain behavior [30]. That is, it comes from the fact that an individual aims to minimize the damage if one needs to make a decision facing a threat that may be caused from an individual's certain behavior [37]. When Bauer [30] first presented this theory, some researches started to multidimensionally deal with elements of perceived risks [15]. Researchers categorized and suggested types of risks of their own professions including performance risk, financial risk, time risk, functional risk, social risk, general risk, etc. Recently, many researchers started to apply the theory of perceived risk into new technology or online service research.

Featherman and Pavlou (2003) integrated TAM model with perceived risk to look for the influence that perceived poses to accepting e-services. The result is the following. People who perceive that using e-transaction service is convenient showed low level of perceived risk, and those with lower perceived risk perceived that e-transaction service is useful and appeared to be willing to continuously accept new e-transaction services [9].

Unlike previous researches, Shin Dong Hee (2008), in the research for acts of transaction with cryptocurrencies at Web 2.0 community, stated the reason why perceived risk was only posing a minor influence over the intention of purchase was because level of perception for risk reduced by the increase of trust as more users gradually started to make purchase using Web 2.0 [32].

We conducted this research by using the perceived risk theory in order to confirm how the perception in Korea, China, and Vietnam that the recent unstable price of cryptocurrencies including Bitcoin and their nature being exchanged in a virtual world pose influences on the intention of usage for cryptocurrencies.

2.4 Theory of Cultural Differences by Countries

Hofstede's cultural comparison theory is the model being cited the most for researching cultures among countries. Hofstede [13] suggested that cultural differences should be compared in six aspects which include power distance, uncertainty avoidance, individualism vs collectivism, masculinity, feminity, long-term orientation vs short-term orientation, and indulgence vs restraint [13].

Power Distance Index (PDI) is a metric for accepting unequal distribution of authorities by small organizational members with small authorities in organizations and groups. Individuals tend to be submissive to those with higher authorities because individuals are influenced by the authoritative order and the mechanism or authority within groups with high power distance [13].

Individualism vs Collectivism (IDV) is the degree of individuals being integrated into organizations. In the individualism society, humans are focused on their own value and greed in terms of their human relationship and they make profits through their given efforts. In the collectivism society, people show mutual interest to group members as well as loyalty and trust to groups [13].

Masculinity vs Feminity: Values within masculine society reflect decision, competitiveness, and desire for achievement. Meanwhile, values within feminity society tend to reflect more interests in humbleness, gentleness, harmony, and ethics [13].

Uncertainty Avoidance Index (UAI) is the degree of resistance and avoidance to ambiguity and uncertainty by people within the society. In a society with high UAI, individuals tend to be anxious and intolerable when facing uncertain situations. Meanwhile, in a society with low UAI, such anxiety is resolved [13].

Long term orientation vs Short term orientation; LTO: In the long term-oriented society, people value patience and future rewards. In the short term oriented society, people value respecting tradition, preserving one's honor, and observing one's duties or responsibilities to attain instant results [13].

Indulgence vs Restraint: In the indulging society, unconstrained activities are tolerated, and it tends to be loose and optimistic. Meanwhile, restraint society has stricter social rules and tends to be strict, tight, and pessimistic [13].

3   Research Design

3.1   Research Model and Hypothesis

Prior to this research, we provide the research model shown in Fig. 2 to conduct empirical analysis of elements affecting intention of cryptocurrency usage among nations based on preceding researches we had considered. Furthermore, research hypothesis is created to clarify relationship among each variable of the research model. It provides the operant definition of measuring parameter and composes questionnaires to measure the concept of variables.

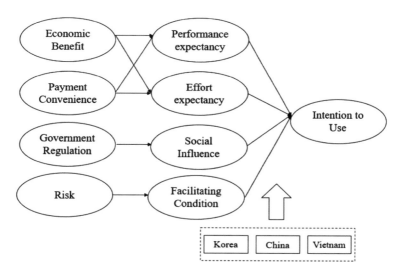

Fig. 2. Research model

Economic feasibility is the degree that assumes that the financial and mental costs required when using new technologies and services are smaller than when using other technologies and services. Garbarino and Edell [8] highlighted that excessively required economic cost may cause negative effects to not only the perceived values for technology and services but also to the selection of usage.

Kotler and Keller [29] claimed that perceived cost of customers is the estimated cost required to use, acquire, access and process products provided at certain markets. This means that temporal, financial, energy, and psychological costs are included, which are critical factors deciding customer's value.

Seong et al. [38], on the research for designing and creating effective micropayment system, asserted that electric currency systems need to require economic feasibility by avoiding unnecessary account creation and collecting or saving unnecessary costs as they build inexpensive currency building mechanism and appropriate security. Shin and Kim [6] stated that using Bitcoin would require less psychological and financial costs than using other electric currencies. They also confirmed that elasticity perceived with economic feasibility had effective influence as it showed static (+) influencing power.

Based on preceding researches above, we expect that economic feasibility may provide positive influences over performance expectancy and effort expectancy. Thus, we provide the following hypothesis.

- H1-1. Economic feasibility will positively (+) affect performance expectancy.
- H1-2. Economic feasibility will positively (+) affect effort expectancy.

Convenience refers to the idea that development of new technology provides convenience to people's daily lives [16]. In this sense, convenience refers to the degree describing how fast and easily trading and payments can be with cryptocurrencies that have effective specific variables [38].

Wendy et al. [26] emphasized in terms of user's utility that electric payment must be transacted with online financial services without causing errors.

Yu et al. [14] asserted convenience was a critical characteristic of electric payment used as a similar meaning to transportability, accessibility, ubiquity, etc. in many researches, showing positive results. Furthermore, Eixom and Todd [2] claimed that convenience influences over intention of usage by using user's attitude toward use of information technology as a medium. Xu and Gutierrez [11] claimed that mobile convenience is a critical element required for a successful mobile commerce.

Shin and Kim [6] stated that payment convenience positively affects to simplicities perceived with elements such as reduction of cash management time, transaction processing, solving inconvenience with spare exchanges, etc. as well as perceived simplicity.

Therefore, researches above assume that payment convenience positively affects performance expectancy and effort expectancy. Thus, we have brought the following hypothesis.

- H2-1. Payment convenience will positively (+) influence performance expectancy
- H2-2. Payment convenience will positively (+) influence effort expectancy

Government regulation policies refer to the methods aiming to instinctively protect many people by regulating or restricting certain activities of some individuals or groups, or methods to enhance general public benefit.

They are justified by external economic effect, loss from natural monopoly, asymmetric information among corporates or between consumers and companies, and prevention of collusion under the monopolistic market structure. Governments aim to activate market competitiveness among corporates with such regulation policies and maximize consumer's benefit. Moreover, in the long term, such regulations are served to strengthen the competitiveness of corresponding industries (Park 2003) [28].

In large, government regulations can be categorized by protective regulations and competitive regulations. Protective regulation refers to the methods used to protect multiple of general citizens or consumers by regulating certain activities of individuals or groups. On the other hand, competitive regulation refers to the method that selects a part of suppliers who can provide materials and services and provides such right. It also is a type of regulation that controls activities of suppliers for the benefit of the public. (R.B Ripley and others, 1986) [31].

Furthermore, preceding researches regarding government regulation had confirmed that most of them positively influenced patents and innovations through technological innovation [1, 17, 21, 23]. However, such researches were focusing on environmental regulations.

The scope of government regulation in this research includes not only legal regulations but also includes social influence and variety. Thus, we have established the following hypothesis.

- H3. Government regulation will negatively (–) influence the social effect.

Accepting technologies are influenced by various factors when new technologies or services show up. In regard to technology or services, how one perceives risks and whether individuals are willing to pay for new services and technologies or to use them significantly influence when making decisions to accept them [44].

Based on the research result about use of illegal copied materials, [43] stated that users perceive various uncertainties which may pose negative influence over attitude.

On a research by Venkatesh et al. (2003) [35], it suggests perceived simplicity, perceived utility, perceived risk, perceived joy, self-effectiveness, and integrity as leading variables and it also expanded the explanatory power of technical acceptance model. As a result, it was confirmed that perceived risk also negatively influences to the intention of using online shopping malls.

Based on these preceding researches, we estimated that perceived risk negatively (–) influences the intention of usage and therefore set up the following hypothesis.

- H4. Risks will negatively (–) influence facilitating conditions.

Performance expectancy represents the degree of trust that use of new information system would enhance work achievement. On the research studying factors influencing the use of Internet banking, Lee et al. [7] claimed that performance expectancy significantly influences the intention of use. On the research regarding clouding computing service, Jung and Nam [4] confirmed that performance expectancy positively influences the behavioral intention. Furthermore, on researches by Mandal and McQueen [5], etc. it was also figured that performance expectancy influence users to accept informative technology. Therefore, we could discover that performance expectancy is the leading variable that greatly influences the behavioral intention. Based on these preceding researches, following hypothesis is provided in this research.

- H5. Performance expectancy will positively (+) influence intention of use
- H6. Effort expectancy will positively (+) influence intention of use
- H7. Social influence will positively (+) influence intention of use
- H8. Facilitating condition will positively (+) influence intention of use

On the research for searching effort achievement and satisfactory level of consumers of Korea and China, Choi Nak Hwan and others [3] stated that there was no difference between Korea and China in terms of information search satisfaction through purchasing knowledge, but it added that feelings such as joy were influenced by cultural factors. Moreover, they added that consumers of China tend to put more effort in seeking information when compared to the consumers of Korea. On the research for influential factors for consumer's purchasing intention for high-tech products by Yoo So Yi [40], it confirmed that influence of subjective rules tend to be more visible among consumers in China than they would in consumers in Korea.

On the research studying Vietnam's telecommunication service product quality and its effect on customer loyalty, Vuong and Kim [36] surveyed Vietnamese users and investigated the telecommunication service product quality's effect on customer satisfaction and the effect on the relationship between customer satisfaction level and customer loyalty. The result was that among service product qualities, calling quality and cellphone, quality of additional service, company image, and attitude and behavior of staff served as the medium for customer satisfaction, posing influence over customer's loyalty.

On the research studying factors influencing intention of accepting mobile payment system in Vietnam, Lee and Kim [22] targeted users in Vietnam and analyzed the customer and systemic characteristic for the intention of using mobile payment. Perceived utility, trust, and compatibility are the critical estimating factors for the intention of using mobile payment. On the research studying for factors influencing the intention of using mobile banking, Uyen [33] targeted Vietnamese customers and the result was that controlled effect such as age, Korean proficiency, and duration of stay in Korea appeared to influence acceptance of mobile banking.

There are a lot of researches for Korea and China regarding the technical acceptance model. However, there is no empirical research regarding Vietnam. Thus, in this research, we referred to researches that had compared various nations, and the following hypothesis is made.

- H9. Respondents of each nation will adjust the relationship among factors that affect the intention of using cryptocurrencies.

3.2 Data Collection and Analysis Method

This research will develop a model by utilizing UTAUT in an attempt to understand the factors that affect the acceptance of cryptocurrencies. Thus, we aim to convey surveys for citizens in Korea, China, and Vietnam asking about the intention of accepting cryptocurrencies. Still, there are mixed perceptions toward cryptocurrencies, and level of perception may differ from each nation. Thus, we provide basic information regarding cryptocurrencies by distributing the respondents with introductory documents about cryptocurrencies before the survey.

Collected samples will go through demographic analysis through SPSS, and AMOS will be used to validate factor analysis, trust and reasonability analysis, and cause and results for the measured variables.

4 Analysis of Actual Proof

4.1 Characteristics of Sample

We conducted frequency analysis to understand the demographic characteristics of the samples to conduct actual proof analysis, and the result is shown in the Table 1 below.

4.2 Exploratory Factor Analysis

We conducted trust and reasonability analysis to validate suitability of research models. We understood the structure among factors by exploring the degree of relationship of inclusive factors for measured variables through exploratory factor analysis (EFA). We conducted the factor analysis with varimax right angle rotation methods by having the original value of factors to be 1.0. We estimated that factor loadings above 0.5 were significant [45]. The result of the analysis indicated that all questions were extracted to 9 components as intended. The result indicated that Cronbach's Alpha values of all measured variables were above 0.7. Therefore, we obtained reliability. Results of EFA and reliability analysis are summarized in Table 2.

4.3 Confirmatory Factor Analysis

We conducted confirmatory factor analysis (CFA) by utilizing AMOS for measured variables obtained from EFA. The result of CFA shows the following research model fit as shown in Table 3.

We confirmed the model fit to be fair and conducted reliability analysis. The result is described in [Table 4]. The measurement model of this research is estimated to have the suitable construct reliability as the construct reliability of all factors satisfied the condition to stay above 0.7 and average variance extracted(AVE) stayed above 0.5 [46].

For distinction reasonability and its result, as shown in [Table 5], the largest one among correlations of latent variables is 0.849. The square of correlation coefficient, in other words, the coefficient of determination is 0.721 (0.849 * 0.849). Therefore, we can assume that it obtained the distinction reasonability since AVE values among all remaining latent variables stay large [46].

4.4 Hypothesis Test

To validate each hypothesis, we conducted route analysis by using AMOS and the results are described in Table 6 and Fig. 3.

4.5 Verification of Control Effect

Depending on respondents of Korea, China, and Vietnam, we conducted control effect analysis to look at the differences with factors that affect the intention of using cryptocurrencies. We analyzed Korea (125), China (43), and Vietnam (40) by dividing respondents by their nationality. To figure whether the difference among these three

Table 1. Demographic characteristics of samples

Type	Frequency (No.)	Ratio (%)	
Nationality	Korea	125	60.1
	Vietnam	43	20.7
	China	40	19.2
Sex	Male	139	66.8
	Female	69	33.2
Age	Under 20	8	3.8
	20's	70	33.7
	30's	41	19.7
	40's	63	30.3
	50's	25	12.0
	60's	1	0.5
Level of Education	High School Graduate or Below	6	2.9
	College Student	39	18.8
	Graduated College	98	47.1
	Graduate School Student	33	15.9
	Graduated Graduate School	32	15.4
Experience	Below 1 year	50	24.0
	Between 1 to 5 years	32	15.4
	Between 5 to 10 years	39	18.8
	Between 10 to 15 years	31	14.9
	Over 15 years	56	26.9
Career	Professional	26	12.5
	Management/Operational	44	21.2
	Retailing/Sales/Service	20	9.6
	Functional Job/ Manual Labor	1	0.5
	Own Business	9	4.3
	Full-time Homemaker	5	2.4
	Student	46	22.1
	Administrative/Technician	49	23.6
	Unemployed	1	0.5
	Others	7	3.4
Annual Income	Below 30 million won	50	24.4
	Between 30 to 50 million won	32	15.4
	Between 50 to 80 million won	39	18.8
	Between 80 to 100 million won	31	14.9
	Over 100 million won	56	26.9
Total	208	100.0	

groups was significant, we validated cross reasonability through Multiple-sample Confirmatory Factor Analysis (MCFA) and validated the control effect through

Table 2. Reliability analysis and EFA

Concept	Component	Cronbach alpha								
	1	2	3	4	5	6	7	8	9	
IU2	.834									0.969
IU5	.818									
IU4	.803									
IU3	.801									
IU1	.762									
EE4		.771								0.93
EE3		.752								
EE1		.650								
EE2		.639								
EE5		.622								
PC1			.778							0.934
PC3			.755							
PC2			.719							
PC4			.670							
RI3				.816						0.901
RI1				.798						
RI2				.794						
SI3					.690					0.915
SI2					.688					
SI4					.626					
SI1					.626					
EB2						.889				0.881
EB1						.815				
EB5						.661				
PE4							.682			0.938
PE2							.670			
PE3							.665			
PE1							.556			
GR4								.917		0.885
GR2								.898		
GR3								.885		
FC3									.648	0.912
FC1									.613	
FC2									.517	

Multiple-group Structural Equation modeling (MSEM). Results are described in [Tables 7, 8, 9, 10, 11]

Looking closely into the analysis result, differences among countries (Korea, China, Vietnam) can be regarded as insignificant as control variables. We clarified that there is

116   K.-J. Jung et al.

Table 3. Model fit of CFA

	Index value		Critical value	
Absolute fit index	Overall model fitness	x^2(CMIN), p	518.001 (p = 0)	p ≦ 0.05 ~ 0.10 (Sample size sensitivity)
		x^2(CMIN)/df	1.682	1.0 ≦ CMIN/df ≦ 3.0
		RMSEA	0.057	≦ 0.05 ~ 0.08
	Model explanatory power	GFI	0.855	≧ 0.8 ~ 0.9
		AGFI	0.809	≧ 0.8 ~ 0.9
		PGFI	0.649	≧ 0.5 ~ 0.6
Incremental fit index	NFI		0.908	≧ 0.8 ~ 0.9
	NNFI(TLI)		0.951	≧ 0.8 ~ 0.9
	CFI		0.960	≧ 0.8 ~ 0.9
Parsimonious fit index	PNFI		0.740	≧ 0.6
	PCFI		0.782	≧ 0.5 ~ 0.6

Table 4. CFA and reliability analysis

	FC	EB	PC	GR	RI	PE	EE	SI	IU
CR	0.915	0.885	0.907	0.886	0.902	0.93	0.915	0.939	0.895
AVE	0.781	0.721	0.767	0.722	0.755	0.816	0.783	0.794	0.743

Table 5. Distinction reasonability of CFA

FC	EB	PC	GR	RI	PE	EE	SI	IU
0.884								
0.497	0.849							
0.668	0.711	0.876						
–0.027	0.020	0.082	0.850					
0.707	0.484	0.535	–0.057	0.869				
0.767	0.569	0.811	0.029	0.676	0.903			
0.700	0.561	0.732	0.071	0.593	0.801	0.885		
0.849	0.465	0.633	0.005	0.636	0.717	0.702	0.891	
0.769	0.529	0.685	–0.006	0.632	0.714	0.716	0.783	0.862

a difference between Korea and Vietnam, although no differences were shown between Korea and China and between China and Vietnam.

After conducting control effect analysis by countries (Korea, Vietnam), for Korea, the hypothesis that government regulation negatively influences society was adopted. However, for Vietnam, the hypothesis that government regulation negatively influences society was rejected. On the contrary, for Korea, the hypothesis that effort expectancy

Table 6. Structural paths assessment and hypotheses test

Result variable		Reason variable	Estimate	S.E.	C.R.	P	Result
PE	<—	EB	−0.149	0.077	−1.939	0.052	Reject
PE	<—	PC	0.986	0.088	11.254	***	Adopt
EE	<—	EB	−0.05	0.077	−0.643	0.52	Reject
EE	<—	PC	0.787	0.085	9.286	***	Adopt
SI	<—	GR	0.006	0.084	0.077	0.939	Reject
FC	<—	RI	0.864	0.08	10.755	***	Adopt
IU	<—	FC	0.187	0.046	4.032	***	Adopt
IU	<—	SI	0.364	0.064	5.71	***	Adopt
IU	<—	EE	0.241	0.08	3.019	0.003	Adopt
IU	<—	PE	0.165	0.075	2.191	0.028	Adopt

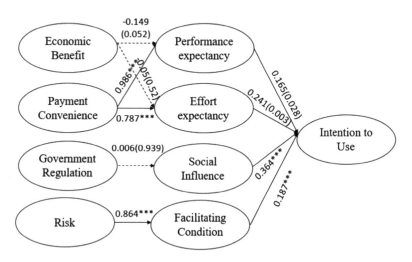

Fig. 3. Verification of structural equation model

Table 7. Result of cross reasonability by nations (Korea, China, Vietnam)

Type	Unconstrained Model	Constraint Model (Structural Weights Model)
χ^2	1963.797 (DF: 984, CMIN/DF: 1.996)	2032.522 (DF:1022, CMIN/DF:1.989)
χ^2 difference	68.725	
P-value of χ^2 difference	0.002	

Table 8. Result of cross reasonability by nations (Korea, Vietnam)

Type	Unconstrained model	Constraint model (Structural weights model)
χ^2	1282.145 (DF:656, CMIN/DF:1.954)	1308.518 (DF:675, CMIN/DF:1.939)
χ^2 difference	26.737	
P-value of χ^2 difference	0.084	

Table 9. Result of cross reasonability by nations (Korea, China)

Type	Unconstrained model	Constraint model (Structural weights model)
χ^2	1054.622 (DF:616, CMIN/DF:1.712)	1099.279 (DF:635, CMIN/DF:1.731)
χ^2 difference	44.657	
P-value of χ^2 difference	0.001	

Table 10. Result of cross reasonability by nations (China, Vietnam)

Type	Unconstrained model	Constraint model (Structural weights model)
χ^2	1068.188 (DF:616, CMIN/DF:1.734)	1103.668 (DF:635, CMIN/DF:1.738)
χ^2 difference	35.48	
P-value of χ^2 difference	0.012	

Table 11. Comparison analysis among Korea, Vietnam

Result variable		Reason variable	Korea (125)			Vietnam (40)		
			Estimate	P	Label	Estimate	P	Label
PE	<—	EB	−0.617	***	Reject	−0.397	0.423	Reject
PE	<—	PC	1.351	***	Adopt	1.659	0.005	Adopt
EE	<—	EB	−0.442	0.003	Reject	−0.356	0.384	Reject
EE	<—	PC	1.177	***	Adopt	1.053	0.024	Adopt
SI	<—	GR	−0.845	***	Adopt	0.072	0.694	Reject
FC	<—	RI	0.824	***	Reject	−0.602	***	Adopt
IU	<—	FC	0.268	***	Adopt	0.217	0.052	Adopt
IU	<—	SI	0.42	***	Adopt	0.247	0.065	Adopt
IU	<—	EE	0.119	0.325	Reject	0.579	0.025	Adopt
IU	<—	PE	0.213	0.106	Reject	0.008	0.968	Reject

positively influences the intention of usage was rejected. However, for Vietnam, the hypothesis that effort expectancy positively influences the intention of usage was adopted.

5   Conclusion

On this research, it conducted an empirical research on factors that affect acceptance of cryptocurrencies. We understood characteristics of cryptocurrencies and designed research model through preceding researches regarding existing UTAUT. Each hypothesis of research models was validated through actual proof analysis, and the following conclusion is given based on the result of validation.

First of all, all hypothesis that expected that economic feasibility would positively affect effort expectancy were rejected (Hypothesis H1-1, H1-2). The reason why economic feasibility is rejected for performance expectancy and effort expectancy is that supply of cryptocurrency is smaller than the supply of credit card. Furthermore, it is estimated that cryptocurrencies would not influence significantly for utility, simplicity, efficiency, and convenience even if cryptocurrencies would cause less usage fee, and distribution and management cost since users are used to using existing currencies and are not well experienced with using cryptocurrencies.

Secondly, the hypothesis (Hypothesis H1-1, H1-2) that payment convenience would positively affect both performance expectancy and effort expectancy were adopted. Payment convenience positively influenced performance expectancy and effort expectancy because people perceived that cash management time reduction from cryptocurrencies and inconvenience from exchanging spares were more convenient and effective than existing currencies.

Third, the hypothesis (H3) which estimated that government regulation would pose a negative influence on social effect was rejected. It is estimated that government regulation, with its system, regulation, rules, and others, influences on economy and subjects of the cryptocurrencies.

Fourth, the hypothesis which estimated that risk would negatively influence facilitating conditions was rejected. This could be due to the idea that cryptocurrencies are not perceived to be riskier than existing currencies. Thus, it is perceived that risk caused while using cryptocurrencies are not perceived to be high regarding issues with cryptocurrency's security such as financial loss and leakage of transaction information and personal information. Moreover, it is estimated that it could be a facilitating condition for the intention of using cryptocurrencies because people in Korea view cryptocurrencies as investment assets and are willing to take a certain degree of risk.

Fifth, the hypothesis(H5) which expected that performance expectancy would positively influence intention of usage, the hypothesis(H6) which estimated that effort expectancy would positively influence intention of usage, the hypothesis(H7) which estimated that social influence will positively affect intention of usage, and the hypothesis(H8) which estimated that faciliating conditions will positively influence intention of usage were adopted. This states that performance expectancy, effort expectancy, social influence and faciliating conditions are believed to be critical elements for potential user's intention of using cryptocurrencies. The result of this research was identical to preceding researches which used UTAUT.

Sixth, we analyzed control effects for each nation (Korea, Vietnam, China). By comparing nations, we figured the difference between Korea and China. For Korea, the hypothesis that government regulation negatively affects social influence was adopted. However, for Vietnam, the hypothesis that government regulation negatively affects social influence was rejected. Uncertainty Avoidance Index of Korea is high and Korean government perceives that there is an on-going uncertainty for cryptocurrencies. Thus, the Korean government advised people not to use cryptocurrencies or restricted people from using them, thereby leading to government regulations which negatively affect social influence. Nevertheless, UAI of Vietnam is low and the sense of uncertainty for cryptocurrencies by Vietnamese government is reduced. Therefore, Vietnamese government neither warns of using cryptocurrencies nor issue regulations restricting the use of cryptocurrencies. Thus, we could figure that it does not negatively affect the social influence regarding the intention of using cryptocurrencies.

Seventh, for Korea, the hypothesis that risk negatively affects faciliating conditions was rejected. However, the hypothesis for Vietnam that risk negatively affects faciliating conditions was adopted. Thus, it is estimated that the market condition of cryptocurrencies in Korea is less risky than the cryptocurrency market in Vietnam. Therefore, risk of cryptocurrencies does not significantly affect faciliating conditions of intention of usage for Koreans. On the other hand, it significantly affects Vietnamese people. Moreover, based on Hofstede's cultural comparison theory, it is believed that the Korean government will manage and develop cryptocurrency technology well since Korea is a long term oriented society. Therefore, risk with cryptocurrencies is believed to be reduced in the future. Nevertheless, in Vietnamese society, since the long term orientation index is lower than that of Korea, it is believed that the there is a high risk for cryptocurrency market only as society may not be capable of estimating the rapid development of cryptocurrency market in the future.

Finally, although the hypothesis for Korea that effort expectancy positively influences intention of usage was rejected, the same hypothesis for Vietnam that effort expectancy positively influences intention of usage was adopted. This states that factors with effort expectancy are not significant and do not influence intention of usage because cryptocurrencies can be used with less effort in Korea than in it would require in Vietnam since Korea is more used to using cryptocurrencies. However, because Vietnam is not well experienced with using cryptocurrencies, effort expectancy factors are critical and believed to influence significantly on the intention of usage.

In regard to views and significance that this research suggests, there are a lot of documents regarding technical characteristics and status presented by research organizations for cryptocurrencies. However, the number of scholastic and empirical research in and outside of Korea is insufficient. In this sense, it is meaningful of this research that it empirically researched about influences on the intention of accepting cryptocurrencies among international society.

Secondly, this research analyzed control effects by dividing respondents by countries. Through comparison among nations, no difference was shown between Korea and China. We also could figure the difference between Korea and Vietnam in terms of the intention of usage. This validated the fact the influences that government regulation gives on the acts of cryptocurrency related markets and economy do not provide negative control effects on cryptocurrencies for Korea and China.

Third, as we compare each nation's acceptance of cryptocurrencies in this research, we suggested a sample by providing UTAUT. This expanded the explanatory power of the model through characteristics of economic feasibility, payment convenience, government regulation and risk from existing technology acceptance model. Moreover, we aim to provide practical sides of lessons for cryptocurrency researches by validating research model and conducting empirical analysis.

Reference

1. Bhatnagar, S., Cohen, M.A.: The impact of environmental regulation on innovation: A panel data study. Unpublished Working Paper. Research Triangle Institute (1997)
2. Wixom, B.H., Todd, P.A.: A theoretical integration of user satisfaction and technology acceptance. Inf. Syst. Res. **16**(1), 85–102 (2005)
3. Choi, N.H., Lee, C.W.: The roles of Brand Image on the Brand-Value-Up. Korean Society for Consumer Culture **9**(3), (2006)
4. Jung, C.H., Nam, S.H.: Cloud computing acceptance at individual level based on extended UTAUT. J. Digital Convergence **12**(1), 287–294 (2014)
5. Mandal, D., McQueen, R.J.: Extending UTAUT to explain social media adoption by microbusinesses. Int. J. Managing Inf. Technol. **4**(4), 1–11 (2012)
6. Shin, D.H., Kim, Y.M.: The factors influencing intention to use bit coin of domestic consumers. J. Korea Contents Assoc. **16**(1), 24–41 (2016)
7. Lee, D.M., Lim, G.H., Jang, S.H.: A comparative analysis on the usage of internet banking users in Korea and China: based on the UTAUT theory. J. Inf. Syst. **19**(4), 111–136 (2010)
8. Garbarino, E.C., Ede, J.A.: Cognitive effort, affect, and choice. J. Consum. Res. **24**(2), 147–158 (1997)
9. Featherman, M.S., Pavlou, P.A.: Predicting e-services adoption: a perceived risk facets perspective. Int. J. Hum Comput Stud. **59**(4), 451–474 (2003)
10. Pieters, G.: Cryptocurrencies as a new global financial asset. In: Allied Social Science Associations (ASSA) 2018 Annual Meeting, Philadelphia, U.S.A., pp. 1–26 (2018)
11. Xu, G., Gutiérrez, J.A.: An exploratory study of killer applications and critical success factors in m-commerce. J. Electron. Commer. Organ. **4**(3), 63–79 (2006)
12. Lee, G.W., Park, Y.J., Park, S.K., Han, S.H., Ryu, S.W.: Technology acceptance model for next generation mobile communication technology of device-to-device. J. Inf. Technol. Archit. **13**(1), 123–137 (2016)
13. Hofstede, G., Hofstede, G.J., Minkov, M.: Cultures and Organizations: Software of the Mind. Revised and Expanded. McGraw-Hill, New York (2010)
14. Yu, H.C., Hsi, K.H., Kuo, P.J.: Electronic payment systems: an analysis and comparison of types. Technol. Soc. **24**(3), 331–347 (2002)
15. Lu, H.P., Hsu, C.L., Hsu, H.Y.: An empirical study of the effect of perceived risk upon intention to use online applications. Inf. Manage. Comput. Secur. **13**(2), 106–120 (2005)
16. Clarke III, I.: Emerging value propositions for m-commerce. J. Bus. Strat. **18**(2), 133–148 (2001)
17. Jaffe, A.B., Palmer, K.: Environmental regulation and innovation: a panel data study. Rev. Econ. Stat. **79**, 610–619 (1997)

18. Jacoby, J., Kaplan, L.B.: The components of perceived risk. In: Venkatesan, M. (ed) SV-Proceedings of the 3rd Annual Conference of the Association for Consumer Research, Chicago, IL, pp. 382–393. Association for Consumer Research (1972)

19. Kim, J.S., Song, T.M.: A study on initial characterization of big data technology acceptance - moderating role of technology user & technology utilizer. J. Korea Contents Assoc. **14**(9), 538–555 (2014)

20. Kim, S.Y.: A Study on the Factors Affecting the Intention to Use Biometrics in Payment Services. Soongsil University Doctoral Thesis (2017)

21. Lanjouw, J.O., Mody, A.: Stimulating Innovation and International Diffusion of Environmentally Responsive Technology. World Bank, Washington DC (1995)

22. Lee, S.T., Kang, W.M., Kim, J.S., Gim, G.Y.: An empirical study on factors affecting customer intention to use mobile payment system in Vietnam. J. IT Serv. Korea **14**(4), 171–184 (2015)

23. Lee, Y.B., Ji, H.J.: Environment regulation, technological innovation, productivity, and their relationship. Korean Assoc. Public Adm. **45**(1), 171–197 (2011)

24. Alshehri, M., Drew, S., Alhussain, T., Alghamd, R.: The effects of website quality on adoption of e-government service: an empirical study applying UTAUT model using SEM. In: 23rd Australasian Conference on Information Systems, Geelong, Australia, pp. 1–13 (2012)

25. Choi, M.J., Lee, D.H., Kim, S.H., Park, H.S., Ahn, H.S.: Impacts of innovative performance through adoption of technology convergence intelligent robot among medium-sized manufacturing firms. J. Digital Convergence **13**(8), 301–313 (2015)

26. Wendy, M.T., Siong, C.C., Binshan, L., Jiat, W.C.: Factors affecting consumers' perception of electronic payment: an empirical analysis. Internet Res. **23**(4), 465–485 (2013)

27. Uyen, P.: An Empirical Study On Factors Affecting Customer Behavioral Intention Towards Mobile Banking. Master thesis of Soongsil University (2015)

28. Park, S.H.: Government regulation and its influence on telecommunication market. Korea Syst. Dyn. Res. **4**(2), 45–69 (2003)

29. Kotler, P., Kelle, L.: Marketing Management. Grada Publishing, Praha (2007)

30. Bauer, R.A.: Consumer behavior as risk taking. In: Hancock, R.S. (ed.) Dynamic Marketing for a Changing world, 1st edn. pp. 389–398. American Marketing Association, Chicago (1960)

31. Ripley, R.B., Franklin, G.A.: Policy Implementation and Bureaucracy. 2nd edn. pp. 72–74. Dorsey, Chicago (1986)

32. Shin, D.H.: Understanding purchasing behaviors in a virtual economy: consumer behavior involving virtual currency in Web 2.0 communities. Interact. Comput. **20**(4–5), 433–446 (2008)

33. Abramova, S., Böhme, R.: Perceived benefit and risk as multidimensional determinants of bitcoin use: a quantitative exploratory study. In: 37th ICIS, Dublin, Ireland, pp. 1–19 (2016)

34. Lee, S.H., Kim, K.S., Kim, S.H.: The impact of intrinsic characteristics of internet technology on technology acceptance. J. Korean Prod. Oper. Manag. Soc. **22**(4), 450–461 (2011)

35. Venkatesh, V., Morris, M.G., Davis, G.B., Davis, F.D.: User acceptance of information technology: toward a unified view. MIS Q. **27**, 425–478 (2003)

36. Vuong, T.B.L.: A Study of the Factors Affecting Intention to Use of Massive Open Online Course in Viet Nam. Master thesis of Soongsil University (2015)

37. Mitchell, V.W.: Consumer perceived risk: conceptualisations and models. Eur. J. Mark. **33**(1/2), 163–195 (1999)

38. Seong, W., Cho, H.S., Cho, H.K., Ham, H.S.: A design and implementation of efficient internet micro-payment system. In: Proceedings of the 27th KISS Spring Conference, pp. 244–246 (2000)
39. Ko, W.H., Shin, K.H., Lee, C.Y.: A study on user behavioral intention of e-procurement system. J. KIIT **13**(11), 167–175 (2015)
40. Lee, Y.S., Park, J.J., Dong, X.F.: A study of behavioral intention to use high-tech products. J. Ind. Econ. Bus. **20**(2) (2007)
41. Tai, Y.M., Ku, Y.C.: Will stock investors use mobile stock trading? A benefit-risk assessment based on a modified UTAUT model. J. Electr. Commer. Res. **14**(1), 67–84 (2013)
42. Wang, Y.M., Wang, Y.S., Yang, Y.F.: Understanding the determinants of RFID adoption in the manufacturing industry. Technol. Forecast. Soc. Change **77**(5), 803–815 (2010)
43. Liao, C., Lin, H.N., Liu, Y.P.: Predicting the use of pirated software: a contingency model integrating perceived risk with the theory of planned behavior. J. Bus. Ethics **91**(2), 237–252 (2010)
44. Holak, S.L., Lehmann, D.R., Sultan, F.: The role of expectations in the adoption of innovative consumer durables: some preliminary evidence. J. Retail. **63**, 243–259 (1987)
45. Lee, H.Y.: Data analysis using SPSS. Cheonglam (2006)
46. Fornell, C., Larcker, D.F.: Structural equation models with unobservable variables and measurement error: algebra and statistics. J. Mark. Res. **18**, 382–388 (1981)

Detection of Replay Attack Traffic in ICS Network

Ki-Seob Hong, Hyo-Bin Kim, Dong-Hyun Kim,
and Jung-Taek Seo$^{(\boxtimes)}$

Department of Information Security Engineering,
Soonchunhyang University, Asan, South Korea
hks0111@gmail.com, sjtgood7@gmail.com,
gyqls1234@naver.com, winer1492@sch.ac.kr

Abstract. The malicious codes and attacks against ICS today are becoming more advanced and intelligent. The security risk for ICS is increasing, and it's becoming more important to secure the cyber safety of ICS from these security threats. Recent ICS not only uses serial communication protocol, but also an Ethernet-based control communication protocol. Malicious codes attacking ICS attempts to imitate the corresponding control protocol to insert malware into the payload for communication, or imitates normal control packets for malicious control or disabling of control devices. Also, multiple presentations exist on the possible scenarios of various cyber attack targeting. However, current IDS/IPS for ICS functions with technology to detect attacks based on a blacklist, and thus cannot detect attacks exhibiting new techniques. In order to solve these problems, there have been recent studies on white list based attack detection technology for practical application on ICS. However, current studies on white list based detection technology utilizes a white list based on IP address, service port number information, etc., and thus cannot be utilized to detect attacks exhibiting a replay pattern or in which only data value is changed inside a normal command. This study suggests a technology that can detect attacks exhibiting a replay pattern against ICS, using white list based detection and machine learning to educate control traffic and apply the results to actual detection.

Keywords: Industrial Control System (ICS) · Network security
Anomaly detection · Replay attack · Machine learning

1 Introduction

Industrial Control System (ICS) is an operating system that integrates a physical system and an IT cyber system. In order to achieve stable operation of the physical system, ICS uses multiple sensors to monitor physical targets and performs actuator control based on current status. The field of ICS includes national infrastructures including energy, power plants, power transmission/transformation, water processing systems, railway systems, airway systems, and shipping. Many other devices are also included in the wider definition of ICS, such as smart home appliances and IoT. The advancement of ICS led to enhanced system operation efficiency and convenience. It's installed in protected areas with limited physical access due to its ability to monitor the current

© Springer Nature Switzerland AG 2019
R. Lee (Ed.): ACIT 2018, SCI 788, pp. 124–136, 2019.
https://doi.org/10.1007/978-3-319-98370-7_10

status using multiple sensors and to control actuator operation based on measured value. However, malicious cyber attacks against IT components of ICS including control system software, network applications and servers are growing more advanced and intelligent with time. Widely known examples of such attacks are the Stuxnet attack that stopped the operation of Iran's uranium extraction facility in 2010, the Duqu attack against the MS Word zeroday weakness in 2011, the Flame attack that infected the Windows system of electricity control in Iran and Middle Eastern countries in 2012. Additional examples include Dragonfly, BlackEnergy, and Industroyer [1–7]. These cases of attacks against ICS prove that ICS is not safe despite its operation in close circuit network format. It's also been confirmed that data manipulation of normal control command transmission can have a malicious impact on physical control devices. The ICS operation environment now attempts the application of vaccines and IDS systems to detect and respond against cyber attacks. However, detection systems using patterns obtained from past analyses cannot detect new forms of attack. There are recent studies aimed at detecting these malicious attacks using anomaly detection, which analyzes normal packets inside ICS to develop white list. However, detection methods using this white list is limited in detecting replay attacks and attacks through data manipulation of normal control commands. Therefore, security technology that can detect these forms of attack must be developed.

In the contents of this study, Sect. 2 discusses the types and features of ICS network structures and protocols. Section 3 discusses the security risk cases posed by the ICS protocol's weaknesses, and the status and limits of ongoing security technology research. Section 4 explains the range and components of problems that this study aims to solve, and analyzes the replay attack communication process that could be considered an ICS network security threat. Section 5 suggests an abnormal traffic detection method using machine learning. Section 6 analyzes the experimental process and results, and Sect. 7 provides the conclusion.

2   Background

This chapter explains the components and features of the typical ICS network structure in order to aid in the understanding of solutions offered in this study. Also, the ICS network protocols used for various purposes are categorized, and their features are investigated and discussed.

2.1   ICS Network Structure

Figure 1 shows the typical ICS Network Structure. It's largely divided into 3 layers, and the IT Network layer consists of a server, firewall, and workstation for internal engineers to perform development & tasks, and it's typically run on TCP communication. The Operation Network layer provides a visual display of a Field device and communicated data using an HMI (Human Machine Interface), enabling the monitoring of ongoing situations. Using the Workstation, an operator can upload and download logic programs from a PLC, and control the PLC by directly sending commands.

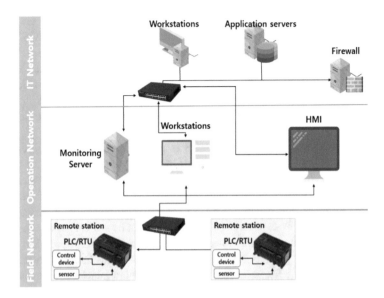

Fig. 1. A typical ICS network structure [8]

Lastly, the Field Network layer consists of a PLC, sensors connected to the PLC, and an actuator.

- **Workstation.** Commonly called Engineering Workstation (EWS), it provides engineers with the ability to manage and control the network PLC and RTU. It's usually provided in the Windows system.
- **Application Server.** An Application Server functions as a buffer between the control bus and the HMI. A PLC cannot withstand large data traffic, and thus the Application Server conducts polling of important data into each PLC and stores it in the database for HMI search.
- **HMI (Human Machine Interface).** An HMI is a device that transforms machine control data into human-friendly form for display. The operator can monitor and control corresponding processes and data.
- **PLC (Programmable Logic Controller), RTU (Remote Terminal Unit).** Directly connected to sensors installed in the process, the PLC transforms the signal from the sensors into digital data recognized by computers and transmits the data to the monitoring system. Recently, PLCs and RTUs provide nearly identical functions and are not categorized separately. An RTU is mainly appropriate for wide ranges and uses wireless communication, while a PLC is mainly used in local control systems in which devices are assigned to one location.
- **Sensor, Actuator.** Sensor/Actuator refers to all devices that are used to monitor, measure and control specific areas of ICS network. For example, actuator may consist of devices such as valve, solenoid, switch, pump, burner and compressor. The sensor measured and inputs value such as temperature, humidity and water volume, and the actuator functions to control motion.

2.2   ICS Network Protocols

ICS Network Protocol provides various protocols based on purpose and environment. Early ICS network only focused on fusibility due to design purpose, and thus there are many protocols that do not consider security elements (confidentiality, integrity, and authentication). However, the introduction of Stuxnet malware led to development and application of protocols with security considerations. Many ICS network protocols apply security at the application level, but several protocol issues were discovered including weakness against simple authentication bypass and descrambling of protocol password routine [9–11]. Table 1 below shows the types and security elements of ICS network protocols that have been frequently used until now.

Table 1. Types and security elements of ICS network protocol

Protocol	Encryption	Authentication	Application level
OPC	OPC UA	OPC UA	OPC UA
DNP3	Secure DNP	Secure DNP	Secure DNP
MODBUS	NO	NO	Modbus
Profinet	NO	NO	Has no security measures of its own
GOOSE	NO	NO	GOOSE
ICCP	NO	NO	Secure ICCP
S7commplus	YES	YES	S7commplus

3   Related Works

3.1   ICS Network Security Threats

The studies on ICS network security threats are currently focused on the Operation Network section and the Field Network's linked section as seen in Fig. 1. Most of the ICS network protocols are designed without security considerations, and field devices can be maliciously controlled using associated weaknesses. Such weaknesses can result in serious risk for the ICS. Klick et al. conducted a study on infecting the internal network of an ICS and PLC connected to the internet using a search engine Shodan, downloading malicious logic code blocks through a weak point of Siemens company's S7comm protocol authentication process [9]. This study confirmed that malicious traffic can be triggered through protocol weakness by penetrating from the external internet network using a simple search, since CPS does not consist of a completely closed network environment. In order to remedy the S7comm's weakness against replay attacks and authentication manipulation, Siemens released a new S7commplus protocol. However, Spenneberg et al. confirmed that the new protocol is still subject to replay attacks through protocol authentication and anti-replay attack function bypass. Also, there have been studies on the dissemination of worm-type malicious code blocks through inter-PLC communication [10]. In response, a new S7commplus protocol was released with updated encryption functionality, but Lei et al. confirmed that normal communication with a PLC can be achieved with a replay attack through decryption of the S7commplus protocol's encryption routine through reverse engineering of TIA Portal, the workstation program that operates and controls the PLC [11]. The above studies confirmed that malicious network traffic can be induced to create abnormal traffic such as Dos attacks or Replay attacks.

3.2 Security Technology and Research of ICS Network

Previous security technology relying on blacklist has shown limited use with the current ICS environment, and thus there are ongoing discussions and actual application of white list based technology [12–14]. The control system the white list is currently being applied to can be categorized into application white list techniques and firewalls using white list rules. Products using techniques in which the application's execution is forbidden include AhnLab's TrustLine, McAfee's Application Control and Industrial Defender's HIDS [15]. Products using other security technology include Tofin's Modbus DPI Firewall, Bayshore Network's SCADA Firewall, and Innominate's Eagle mGuard, and these products provide white list functionality using simple MAC, IP and Port, and supply monitoring functionality through analysis of detailed traffic of ICS network protocols (Modbus, DNP3, ICCP, etc.) [16–18]. A recent study has suggested new-generation IDS, an IDS system customized for ICS networks through stratification of access control white list, protocol-based white list and activity-based rules [19]. However, pattern-based detection methods cannot detect new forms of attacks, and white list based detection methods under ongoing research pose difficulty in detecting replay attacks disguised as normal data. Accordingly, there are diverse ongoing studies related to the anomaly detection of the control system [20–22].

1. Yasakethu and Jiang performed a study on invasion detection through machine learning for SCADA system protection and categorized normal and abnormal data using the OCSVM model [20].
2. Ponomarev and Atkison detected non-permitted remote access to PLC using a telemetry-based intrusion detection system. This study detected invasion using categorizing algorithm that utilizes the following data: time it takes for a client to respond to server message, amount of client's drop packet, amount of server's drop packet, and time between repeated packet transfers in case of packet loss [21].
3. Schuster et al. detected intrusion to the PROFINET protocol from the external environment by using the n-grams model through the selection of source and destination addresses, protocol type, packet type, and packet data as its features, which are continuous and categorical in nature [22].

4 Security Threats Through ICS Network Protocol Analysis

This section determines the experimental subject range and components of this study, which attempts to establish a detection solution against abnormal traffic in the ICS network. And the abnormal communication process of a replay attack, used as a security threat, is analyzed and executed through protocol analysis.

4.1   Subject Range and Components

The subject range is determined as the communication section of Operation Network and Field Network as shown in Fig. 1. Table 2 shows the subjects and corresponding components.

Table 2. Experimental subjects and components

Subject	Component
PLC under analysis	Siemens PLC 1214c AC/DC/Rly, 1214c AC/DC/Rly
PLC firmware version	Version 3.0
Operation system	Windows 7 Ultimate K
Analytic protocol	S7commplus
TIA portal version	Version 11

It is difficult to obtain the dataset of an actual ICS due to corporate and institutional security concerns. Therefore, a testbed was created for experimental purposes, in which a closed network environment with no connection to external internet was established as seen in Fig. 2. The PC functioning as a Workstation was installed with TIA 11 software, which allows operation/control of the PLC and download/upload of the logic program, for direct communication with the PLC.

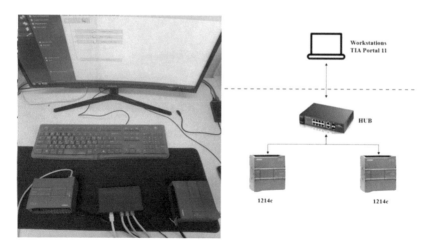

Fig. 2. Testbed and network arrangement

4.2   S7commplus Protocol Replay Attack

The S7commplus protocol operates on Ethernet, and each PLC is assigned an IP address to communicate with each other via a No. 102 port. For the replay attack, the communication process is analyzed through packet capture. Communication between PC(TIA Portal) and PLC, and between each PLC is attempted by 3-way-handshade connection as shown in Fig. 3, and the PLC's random 1 byte is transmitted to the PC. The PC adds 0×80 value to the random byte received, and the result is sent through the 24th and 29th data payload of the S7commplus request. The PLC confirms the response value received and forms a session to conduct operation, either to stop or start the PLC by sending the request along with the calculated byte value to initiate diverse functions including PLD Stop, Start, or logic program upload/download.

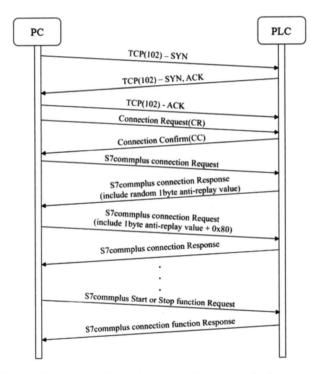

Fig. 3. S7commplus Protocol replay packet communication process

5   Suggestion on Abnormal Traffic Detection Solution

5.1   Replay Attack Traffic Detection Scheme Using Autoencoder Model

In order to detect replay attack traffic, this study generates S7commplus protocol replay attack traffic by establishing a testbed as in Fig. 2 in order to analyze the difference from normal traffic. A general neural network may be used since normal and abnormal traffic is labeled in this case, but in a real ICS environment, it's nearly impossible to differentiate abnormal traffic and obtain labeled data. Therefore, an unsupervised learning algorithm is selected rather than supervised learning that assumed labeling of data. For the model, this study suggests AutoEncoder to detect abnormal traffic detection. Data not learned through AutoEncoder exhibits different values compared to those entered by decoder and original data. Based on the model learned with normal traffic data, this method detected normal/abnormal traffic through the difference in result and input values when abnormal traffic value enters. Prior to subjecting the model to learning, it's important to select differentiated input data to categorize normal and abnormal traffic data. A replay attack is characterized by capturing normal traffic activity and sending it directly to the target, making it very difficult to detect. Security techniques to prevent replay attacks include time synchronization or nonce value utilization. However, the nonce value can be calculated in the S7commplus protocol to initiate a replay attack, and it's extremely difficult to detect the difference between normal and abnormal traffic. Also, if a PC connected to a PLC is already infected with malignant code in a control network, a replay attack may be initiated to the PLC using a malignant script. This case is particularly difficult to detect since the attacking subject will possess the same host pc name, port and IP. In this study, a testbed was prepared as seen in Fig. 2 to analyze normal and abnormal traffic in Start/Stop activity of the PLC, and the difference in inter packet arrive time was discovered as a result. For start and stop functions, 18 uniform sequences of request packets is generated per 1 flow. Due to the difference is size among each of the 18 packets, the inter-packet arrival time is different among each. And it's been discovered that the normal PLC start/stop function through the TIA Portal exhibits a different inter-packet arrive time compared to a packet generated from the PLC start/stop function generalized through replay attack with a malicious script. Tables 3 and 4 below show the average value of the above mentioned time value.

Table 3. Comparison of average inter packet arrive time between normal/abnormal start function

Start function sequence packet	Normal average inter-packet arrival time (second)	Abnormal average inter-packet arrival time (second)
Request packet1	0.010748	0.094736
Request packet2	0.009148	0.161236
Request packet3	0.00935	0.000147
Request packet4	0.001457	2.64242E−05
Request packet5	0.00915	3.16364E−05
Request packet6	0.010219	3.24545E−05
Request packet7	0.029905	2.83939E−05
Request packet8	0.008563	2.76061E−05
Request packet9	0.015661	2.37273E−05
Request packet10	0.008952	2.55455E−05
Request packet11	0.0126	2.55455E−05
Request packet12	0.029052	0.00014
Request packet13	0.027653	0.00012
Request packet14	0.01504	3.08182E−05
Request packet15	0.010019	2.6303E−05
Request packet16	0.011326	2.75455E−05
Request packet17	0.002146	2.74545E−05
Request packet18	0.025697	2.55455E−05

Table 4. Comparison of average inter-packet arrival time between normal/abnormal stop function

Stop function sequence packet	Normal average inter packet arrive time (second)	Abnormal average inter packet arrive time (second)
Request packet1	0.001469	0.000129
Request packet2	0.001211	2.82703E−05
Request packet3	0.001793	3.25135E−05
Request packet4	0.001154	3.90541E−05
Request packet5	0.029702	3.54324E−05
Request packet6	0.00141	3.35135E−05
Request packet7	0.001089	3.36486E−05
Request packet8	0.000899	3.83514E−05
Request packet9	0.076845	3.71081E−05
Request packet10	0.013637	3.81351E−05
Request packet11	0.001599	3.36757E−05
Request packet12	0.001221	2.94054E−05
Request packet13	0.017105	2.80541E−05
Request packet14	0.029586	2.98378E−05
Request packet15	0.001795	2.62162E−05
Request packet16	0.001258	3.24324E−05
Request packet17	0.004719	2.85676E−05
Request packet18	0.008088	2.79189E−05

The 18 inter-packet arrival time values from normal traffic are selected from Tables 3 and 4 as input data as shown in Fig. 4, and learning is induced via the Autoencoder model. The model learned through normal traffic conducts the detection of abnormal traffic with the difference in result value when the input data from abnormal traffic enters.

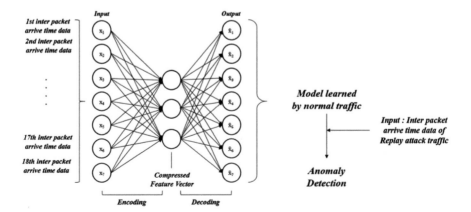

Fig. 4. Replay attack traffic detection model scheme using Autoencoder

6 Experiment

6.1 Experimental Environment and Components

In order to compose the learned model suggested in Sect. 5, the Tensorflow, an open machine learning system released by Google, was utilized. To collect the input data, this study collected 351 packet flows corresponding to stop function and 360 packet flows corresponding to start function through the TIA portal for 1 h, and the learning was conducted based on this data. As shown in Fig. 5, normal data among the collected data are divided into learning-purpose, test-purpose, and verification-purpose by 60%, 20%, and 20%, respectively. The abnormal data consisting of 33 stop function flows and 38 start function flows were divided into 50%/50% for test purpose and verification purpose. For this step, the data was first divided into normal and abnormal data sets. Test-purpose data consisted of both normal and abnormal traffic data. The normal and abnormal traffic flows were mixed through a shuffling process and the mixture was utilized as input for the model, which underwent learning through normal traffic data, to verify the existence of objective and accurate detection.

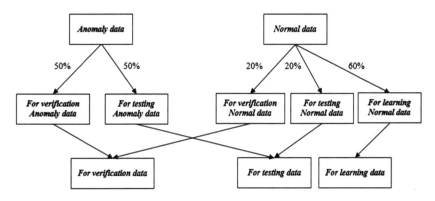

Fig. 5. Replay attack traffic detection model scheme using Autoencoder

6.2 Experiment Results

The total data collected for verification purpose were 89 start function and 88 stop function inter-packet arrival times, and the amount of abnormal traffic was 19 and 16 packet flows, respectively. It's confirmed that 100% of the replay attack's abnormal traffic is detected by threshold value. For the detection of intrusion from the external environment, existing studies mainly selected static or distinguishable features such as ICS protocol packet type, timestamp, source, destination, addresses, and packet loss to establish diverse models of the categorization technique [20–22]. This study mainly focused on replay attack traffic detection that occurs inside a PC infected with external malicious code. There are limits to detecting replay attacks using features selected from the abovementioned studies. Through experimentation, this study was able to detect a difference in inter-packet arrival time between normal packet and replay attack packet and selected this value as a feature. Furthermore, this study was able to categorize replay attack traffic through deviation from normal values using the unsupervised learning model Autoencoder (Table 5).

Table 5. Detection of abnormal traffic using Autoencoder

Threshold	Amount of abnormal traffic (in packets) Detected in start function	Amount of abnormal traffic (in packets) Detected in stop function
1.1	19	16
1.0	19	16
0.9	19	16
0.8	19	16

7 Conclusion

This study was conducted to overcome the limits of white list as discussed in Sect. 3, and its experiment was conducted by establishing a testbed to analyze normal traffic and a replay attack's abnormal traffic. It was confirmed that the replay attack's abnormal traffic exhibits different inter-packet arrival times compared to normal traffic. The discrepancies found in the experimental results are analyzed to be the result of the difference in traffic transmission speed to the PLC through the TIA Portal's control command transmission logic compared to the transmission speed of a malicious script's logic carrying out a replay attack, and also due to the difference in the PC and network environment of the ICS.

Using Autoencoder, a type of unsupervised learning mode, this study confirmed that all of a replay attack's abnormal traffic is able to be detected by a model that learned the inter-packet arrive time of normal traffic by detecting the difference in output value induced by abnormal traffic's input. The analysis of the experimental results also confirmed the possibility of detecting a replay attack's abnormal traffic in a closed network environment such as an ICS through machine learning. In order to effectively detect various cyber-attacks including replay attacks to secure cyber stability in ICS environments, a wide range of detection technologies must be utilized including traditional white list, hybrid white list under ongoing research, and the detection of abnormal traffic detection through machine learning as suggested in this study.

However, this study exhibits limits in the experimental results due to the difficulty in collecting large datasets seen in actual ICSs conducting various functions, which led to limited learning data through testbed preparation. Future studies are needed on real-time detection of abnormal traffic through machine learning for practical purposes, and effective machine learning algorithms to study large-volume traffic data of actual ICSs that conduct various functions.

Acknowledgement. This research was supported by the Research Program of the Korea Institute of Energy Technology Evaluation and Planning (KETEP) Institute of Korea. (No. 20162220200010) and the Soonchunhyang University Research Fund.

References

1. Falliere, N., O Murchu, L., Chien, E.: W32.Stuxnet Dossier, Symantec, version 1.3 edition, November 2010
2. Virvilis, N., Gritzalis, D.: The big four - what we did wrong in advanced persistent threat detection? In: IEEE Availability, Reliability and Security (ARES), pp. 248–254, September 2013
3. Bencsath, B., Pek, G., Buttyan, L., Felegyhazi, M.: The cousins of Stuxnet: Duqu, Flame, and Gauss. Proc. Future Internet **4**(4), 971–1003 (2012)
4. Piggin, R.: Critical infrastructure under attack. ITNOW **56**(4), 30–33 (2014)
5. Khan, R., Maynard, P., McLaughlin, K., Laverty, D., Sezer, S.: Threat analysis of BlackEnergy malware for synchrophasor based real-time control and monitoring in smart grid. In: Proceedings of 4th international symposium on ICS SCADA cyber security research (ICS-CSR), pp. 53–63, August 2016
6. Cherepanov, A.: Win32/INDUSTROYER-a new threat for industrial control systems, Technical report (2017). https://www.welivesecurity.com/wpcontent/uploads/2017/06/Win32_Industroyer.pdf
7. Nazir, S., Patel, S., Patel, D.: Assessing and augmenting SCADA cyber security: a survey of techniques. Comput. Secur. **70**, 436–454 (2017)
8. Maglaras, L.A., Jiang, J., Cruz, T.J.: Integrated OCSVM mechanism for intrusion detection in SCADA systems. IET Electron. Lett. **50**, 1935–1936 (2014)
9. Klick, J., Lau, S., Marzin, D., Malchow, J.-O., Roth, V.: Internet-facing PLCs - a new back orifice. In: Blackhat USA 2015, Las Vegas, USA (2015)
10. Spenneberg, R., Brüggemann, M., Schwartke, H.: PLC-blaster: a worm living solely in the PLC. In: Blackhat ASIA 2016, Singapore (2016)
11. Lei, C., Donghong, L., Liang, M.: The spear to break the security wall of S7CommPlus. In: Blackhat USA 2017, Las Vegas USA (2017)
12. Ginter, A.: An analysis of Whitelisting security solutions and their applicability in control systems. In: SCADA Security Scientific Symposium (S4), Miami, USA, January 2010
13. Yoon, J., Kim, W., Seo, J.: Study on technology requirement using the technological trend of security products concerning industrial control system. J. Korea Inst. Inform. Secur. Crytol. **22**(5), 22–26 (2012)
14. Barbosa, R.R.R., Sadre, R., Pras, A.: Flow whitelisting in SCADA networks. Int. J. Crit. Infrastruct. Protect. **6**(3), 150–158 (2013)
15. Yoo, H., Yun, J.-H., Shon, T.: Whitelist-based anomaly detection for industrial control system security. J. KICS **38**(08), 641–653 (2013)
16. The Tofino security appliance website (2015). http://www.tofinosecurity.com/products
17. The innominate security technologies mGuard website (2015). http://www.innominate.com/en/products
18. Kim, B.K., Kang, D.H., Na, J.C., Chung, T.M.: Abnormal traffic filtering mechanism for protecting ICS networks. In: 2016 18th International Conference on Advanced Communication Technology (ICACT), pp. 436–440. IEEE, January 2016
19. Yang, Y., et al.: Multiattribute SCADAspecific intrusion detection system for power networks. IEEE Trans. Power Deliv. **29**(3), 1092–1102 (2014)
20. Yasakethu, S.L.P., Jiang, J.: Intrusion detection via machine learning for SCADA system protection. In: Proceedings of the 1st International Symposium on ICS & SCADA Cyber Security Research 2013, pp. 101–105, 16–17 September 2013, Leicester, UK (2013)

21. Ponomarev, S., Atkison, T.: Industrial control system network intrusion detection by telemetry analysis. IEEE Trans. Dependable Secure Comput. **13**(2), 252–260 (2016)
22. Schuster, F., Paul, A., König, H.: Towards learning normality for anomaly detection in industrial control networks. In: Doyen, G., Waldburger, M., Čeleda, P., Sperotto, A., Stiller, B. (eds.) Emerging Management Mechanisms for the Future Internet. AIMS 2013. LNCS, vol. 7943. Springer, Heidelberg (2013). https://doi.org/10.1007/978-3-642-38998-6_8

CipherBit192: Encryption Technique for Securing Data

Benedicto B. Balilo Jr.[1], Bobby D. Gerardo[2],
and Yungcheol Byun[3(✉)]

[1] CSIT Department, Bicol University Legazpi City, Albay, Philippines
benedicto.balilojr@gmail.com
[2] Institute of ICT, West Visayas State University, Lapaz Iloilo City, Philippines
bgerardo@wvsu.edu.ph
[3] Department of Computer Engineering, Jeju National University, Jeju, Korea
ycb@jejunu.ac.kr

Abstract. Cryptography is an important component in securing and protecting sensitive and confidential data. It is a technique of encrypting data, hiding the process and produced a ciphered text. The Data Encryption Standard (DES) algorithm is considerably known for its branded encryption/decryption technique. To explore the advantage of its recognized cycle process and ciphertext representation. This paper presents the new algorithm based on mixed technique by substituting reverse scheme, swapping and bit-manipulation. This encryption technique is a concept that combines selected cryptographic technique that can reduce and save time processing encryption operation, offers simplicity, and extended mixed encryption key generation.

Keywords: Encryption · Decryption · Cryptography · Securing data

1 Introduction

The growth of the internet provides opportunities and communication link between the tens of billions of people and increasingly took advantage for social media, commerce, marketing, and many others, thus, security becomes the central core of issue. Cryptography is the technique of processing and transmitting data into a form where only the intended recipient can read and process it. One aspect of securing communication is the conversion of the data in a form called ciphertext, which both the sender and recipient know only the key generated.

The term encryption comes from the Greek word kryptos, meaning hidden or secret. Cryptography is a method of storing and transmitting data in a form known only to the reader. Cryptology is the study of both cryptography and cryptanalysis. Today's cryptosystems are divided into two categories: symmetric and asymmetric. Symmetric crypto systems use the same key (the secret key) to encrypt and decrypt a message, and asymmetric cryptosystems use one key (the public key) to encrypt a message and a different key (the private key) to decrypt it and all of today's algorithms [1].

© Springer Nature Switzerland AG 2019
R. Lee (Ed.): ACIT 2018, SCI 788, pp. 137–148, 2019.
https://doi.org/10.1007/978-3-319-98370-7_11

Conventional encryption is a classical technique which was used prior to the development of public key encryption. Cryptography are classified into three independent dimensions namely: the type of operation used for transforming plaintext to ciphertext (by substitution and transposition), the number of keys used, and the way in which plaintext is processed (block or stream). The operation of plaintext message can be done in many different ways like substitution techniques, Caesar cipher, playfair, hill cipher and many others. But, with the introduction of modern encryption techniques like block cipher based on Feistel Cipher structure cryptographic support offers much comparison in any ways [2].

The Data Encryption Standard (DES) is a symmetric-key block cipher that uses 16 rounds Feistel structure. It's block size is 64-bit. It was used by industries in embedded systems, SIM cards, smart cards and network devices that require encryption. The DES algorithm encrypts data after a number of rounds including hashing, permutations, and shifting operations. It has been a popular secret key encryption and is used in many commercial and financial applications. However, its key size is too small by current standards and its entire 56-bit key space can be searched in approximately 22 h. Though, it satisfies the avalanche effect and completeness properties of block cipher making it very strong. However, over the years, cryptanalysis found some weaknesses in DES when key selected are weak keys which can be avoided. DES was succeeded by Advanced Encryption Standard (AES) algorithm defined in Federal Information Processing Standard (FIPS) 192 for Federal use in the US [23] which is more secure and became the choice of various security services in numerous applications like in banking to secure online transactions, cloud storage system, serial communication (universal asynchronous receiver-transmitter (UARL) for secure transfer of data which are RTL (VHDL) [25]) and others [3, 22]. The AES developed by Rijndael was been made using Verilog code which can be easily implemented on Field Programmable Gate Arrays (FPGA) which composed of three main parts: cipher, inverse cipher and Key Expansion. The cipher converts the data to plaintext. Key expansion generates the key schedule used in cipher and inverse cipher procedure. The inverse cipher composed of number of rounds, during the execution of the algorithm these rounds are performed which composed of different byte-oriented transformations: Sub Bytes, Shift Rows, Mix columns and Add Round Key [24].

Due to DES branded and recognized ciphertext representation, a number of comparative researches on different algorithms have been conducted to improve its performance and security through manipulating the cycle process. Others used several factors to measure the performance of the algorithm such as key length, cipher type, block size, security, possible keys, possible ASCII printable character keys and time required to check all possible keys at 50 billion keys per second [20]. In some researches, these factors are used to analyze the performance of the algorithm in terms of authentication, flexibility, reliability, robustness, scalability and security to highlight the weaknesses of the algorithm or even to identify the strength and limitations as applied to an application [21].

The application of substitution technique before the DES performed its process requires the intruder to break both the substitution and original DES algorithm [17]. The Double DES (2DES) algorithm is similar to DES but with repeated process on selected cycle. The DES was applied twice using two keys K1 and K2 which leads to

112 bit key [18]. The 2DES is much more secure than the original DES algorithm but produced smaller keys, this leads to the introduction of Triple DES Algorithm (3DES). The 3DES has 168-bit key length larger than the previous algorithm [19].

DES Message Digest Computation uses a DES variant as a one-way hash function. It 1994, DMDC was introduced to compute the 18-bit authentication data internet security. The DMDC hash function generates message digests with variable sizes – 18, 32, 64 or 128 bits. The message to be signed is first divided into a sequence of 64-bit blocks. This scheme is appropriate for the USE OF DIGITAL SIGNATURES, to improve internet security. The Triple DES (3DES) was approved through 2030 for sensitive government information. It performs the three iterations of the DES algorithm, that is, if keying option number one is chosen, a different key is used each time to increase the key length to 168 bits. 3DES encryption is obviously slower than plain DES.

The Playfair cipher was the first practical digraphic substitution cipher invented by Charles Wheatstone in 1854 but later named after Lord Playfair who promoted the use of cipher. The technique encrypts pairs of letters (digraphs), instead of single letters as in the simple substitution cipher [4].

The Blowfish was designed by 1993 by Bruce Schneier. It is a 16-round Feistel cipher which has a 64-bit block size and a variable-length key length from 32 bits to 448 bits. The S-boxes accept 8-bit input and produce 32-bit output. The function splits the 32-bit input into four eight-bit quarters, and uses the quarters as input to the S-boxes. The outputs are added modulo 232 and XORed to produce the final 32-bit output. Since Blowfish is a Feistel network, it can be inverted simply by XORing P17 and P18 to the ciphertext block, then using the P-entries in reverse order [5]. There are examination conducted to Blowfish including the identification of weak keys, the class keys that can be detected—although not broken-in (variants of 14 rounds or less), and second-order differential attack on 4-round that cannot be extended to more rounds. But, Blowfish is much faster algorithm than DES and IDEA [6]. Blowfish are used in products such as in file encryption and wipe utility for all Win32 systems, file browser, job automation, auto password confirmation, password manager for windows, and in word processor incorporating text encryption [7].

In Table 1, shows the comparison of some cryptographic functions and their respective key length. The comparison of each algorithm depends on the key characteristics and properties how to secure plaintext in number of cycles and bits manipulation.

The SHA was designed by NIST & NSA in 1993, revised 1995 as SHA-1. It was used by US as standard use with DSA signature scheme which produces 160-bit hash values (now the generally preferred hash algorithm). SHA-1 was based on design of MD4 with key differences with pad message length of 448 mod 512 which append a 64-bit length value to message. It expanded to 16 words into 80 words by mixing and shifting [8].

The bit manipulation works on manipulating the bits or other pieces shorter than a word. It reduces the need to loop over the structure and speed up the process. Some of the tasks that require bit manipulation include low-level device, error detection and correction algorithms, data compression, encryption algorithms, and optimization [9]. Playfair combined with bit manipulation provides an interesting key generation along with the application of swapping, shifting and reversal of text.

Table 1. Some Cryptographic functions and their key length

Cryptographic function	Key lengths		Initialization vector	
	In Bytes	In Bits	In Bytes	In Bits
AES	16, 24 or 32	128, 192 or 256	16	128
DES	1 to 8 bytes	8 to 64	16	128
TRIPLEDES	1 to 24	8 to 192	16	128
BLOWFISH	1 to 56	8 to 448	16	128
TWOFISH	1 to 32	8 to 256	32	256
RIJNDAEL-256	1 to 32	8 to 256	64	512
R4	1 to 256	8 to 2048	–	–
RC5	1 to 2040	8 to 16320	32	256
SERPENT	1 to 32	8 to 256	32	256

The application of reversing text has long been used like in Julius Ceasar cipher of which deciphering is done in reverse with a right shift of 3. Reverse Encryption Algorithm (REA) because of its simplicity and efficiency it limits the added time cost for encryption and decryption. It takes a variable-length key, it added the keys to the text in the encipherment and removed the keys from the text in the decipherment, and executed divide operation on the text by 4 in the encipherment and executed multiple operation on the text by 4 in the decipherment. It used the divide operation by 4 on the text to narrow the range domain of the ASCII code table at converting the text [11].

In this paper, the encryption algorithm is a block cipher based on the combination of reverse cipher, swapping and shifting bit manipulation which operates on blocks of data. The value must be a combination of numbers and alpha character symbols. The block size varies on different cycles each generates from plaintext to bit conversion. [10] introduced the transposition technique algorithms—reverse transposition cipher and odd-even transposition technique. The algorithm was based on the number of words, lines or paragraphs. It randomized the key which makes the process of cryptography more secure.

In [12] proposed a new encryption scheme (Chaotic Order Preserving Encryption (COPE)). It hides the order of the encrypted values by changing the order of buckets in the plaintext domain. It is secure against known plaintext attack. However, COPE can be used just on trusted server where the encryption keys are used to perform many queries such as join and range queries. The overhead of range queries over encrypted database is much higher than the overhead of range queries over plaintext database. In addition, it uses many keys to change the order of buckets and in some cases that may lead to have duplicated values. Another drawback in COPE is the encryption and decryption cost. That is because of the computation complexity to randomize the buckets and assign the correct order within each bucket.

The bucketing approach [13, 14, 16, 17] is dividing the plaintext domain into many partitions (buckets). The encrypted database in the bucketing approach is augmented with additional information (the index of attributes), thereby allowing query processing to some extent at the server without endangering data privacy. The encrypted database in the bucketing approach contains etuples and corresponding bucket-ids (where many plaintext values are indexed to same bucket-id). In this scheme, executing a query over the encrypted database is based on the index of attributes. The result of this query is a superset of records containing false positive tuples. These false hits must be removed in a post filtering process after etuples returned by the query are decrypted. Because only the bucket-id is used in a join operation, filtering can be complex, especially when random mapping is used to assign bucket-ids rather than order preserving mapping. In bucketing, the projection operation is not implemented over the encrypted database, because a row level encryption is used.

2 The Proposed Work

The algorithm used the concept of reverse, swapping and shifting of plaintext to its binary equivalent. Series of steps have been tested and implemented to produce the encrypted value. The 32-bits representation in each block carries the plaintext in Byte before converting it to its equivalent ASCII representation. As a final result, the plaintext shall be transformed into its equivalent encrypted values.

2.1 CipherText Structure

The ciphertext structure defines the organization of the proposed algorithm that each round performed different operations which composed of different bit-byte manipulation: byte splitting, block switch, shift rows and mix rows. Given the value of $M = 12$ characters, the value of M is a combination of string of characters and number. If the value of M is less than the given length of the text a separator symbol shall be appended to the actual text–plaintext. The reversed plaintext M follows the cycle of operations.

Figure 1 shows the structure given the 96-bit input and produced a 24-Byte ciphertext schedule. The encryption was divided into four (4) groups combining simple transposition, byte splitting, swapping and shift rows to the plaintext. The swapping of L1 and L2 over R1 and R2 position (vice versa) shows a twist technique where characters are converted into 8-Bit binary representation and perform reversal of the bits. This offers simplicity yet complex first layer method to plaintext before entering the shift rows layer operation.

The plaintext composed of two (2) divisions (L-R), the first division used the shift rows representation operation while the remaining division performed the inverse transposition of bits. Thus it combines the results produced by R1 and R2 over L1 and L2 swapping operations into an encrypted 24-Byte schedule.

2.2 Cipher Algorithm

The encryption pseudocode of the encryption and decryption process depicts the concept of the proposed algorithm. The reversal and swapping of grouped bits on specified cycle operation allows the process to change the position of bits thereby generating dissimilar BCD 8-4-2-1 values. The BCD 8-4-2-1 code is a coding scheme

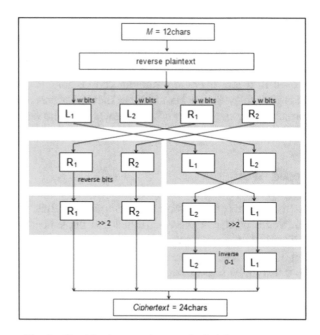

Fig. 1. Combined encryption standard ciphertext structure

represented by a group of 4 binary digits or bits which described the code equivalent to decimal value [26].

In the encryption process, the plaintext is a combination of 12-Bytes or characters–including numbers (Line 1). In the event that the plaintext obtained less than the required length a separator will be added in the plaintext (Line 3).

Encryption

   1. plaintext
   2. if plaintext less than required length
   3.   add separator
   4. end
   5. randomized text
   6. reverse plaintext
   7. do looping
   8.   divide plaintext
   9.   swap L1, L2 over R1 and R2
  10.   right shift 2 R1 and R2
  11.   swap L1 and L2
  12.   reverse L1 and L2
  13. end
  14. encrypted text

The decryption process allows the encrypted text to reveal the original text or decrypted value. The decrypted text was divided into 48-bits allowing the process determine the sequence and pattern of decrypting the text.

The separator served as the parameter limit which determines if the decrypted value carries a full length ciphertext or less than the defined length. In the cycle process, the decrypted text was divided into 48-bits representation to provide easy identification of the L and R division (Line 3). This draws the determination of L1, L2 and R1 and R2 block (Line 4). The separation of blocks follows the inversion process, shift rows and swapping operation leading to the decryption of the encrypted text (Line 12).

Decryption

```
1. encrypted text
2.    do looping
3.       divide encrypted text
4.       read L1 and L2
5.       inverse L1 and L2
6.       left shift 2 L1, L2, R1 and R2
7.       reverse bits R1 and R2
8.       swap L1 and L2
9.       reverse L1, L2, R1, and R2
10. end
11. locate/remove separator
12. decrypted text
```

3   Results and Discussions

3.1   Encryption

After analysis and evaluation to the new encryption algorithm, it offers simplicity yet complicated cycle operation, new concept, and extended mixed encryption key generation. This provides limited key cycles/round but offers high encrypted values. The algorithm is a block cipher that can be used for various encryption applications like securing online transactions. The algorithm divides the text into two (2) divisions making each division perform different operation in each cycle level.

Some algorithms implemented a division of two or three blocks to plaintext, the proposed algorithm offers four blocks of 24-bits. For purposes of presentation, let us assume that the text "bicolunivers" be served as plaintext used in the discussion. Given the 12-Byte key length, the first step includes division and reversal of text to form the 6-Byte group as shown in Fig. 2.

The reversed plaintext was divided into two blocks and performs swapping from left to right sequence. Same with other transposition of text, the algorithm offers swapping of blocks and conversion of the same to 4-bit BCD schedule. As shown in Figs. 3 and 4, the right block was transposed and divided into 3-Byte length making it 2-blocks.

144   B. B. Balilo Jr. et al.

1st 6 characters	2nd 6 characters
srevin	ulocib

Fig. 2. Transposition and division of text to 2-block

1-3	3-6
ulo	cib
01110101011011000110111	011000110110100101100010
===================	===================
111101100011011010101110	010001101001011011000110
f636ae	4696c6

Fig. 3. Left side (LS) transposition and division of text

7-9	10-12
sre	vin
011100110111001001100101	011101100110100101101110
===================	===================
011100110111001001100101	011101100110100101101110
737265	76696e

Fig. 4. Right side (RS) transposition and division of text

The LS performed BitShift to the right by two and obtain the equivalent hexadecimal value. While, RS performed swapping of the fourth over the third block then combine the result. These combinations directed the operation to perform 2 BitShift to the right and obtain the hexadecimal value.

The swapping of values from one block to the other together with the combination of BitShift along with the inclusion of inverting 0's and 1's offers assurance and complexity, as shown in Fig. 5. Through the BCD 8421 code representation, the

swap cluster (10-12 move to 7-9) then and >> 2	
110110011010010110111001	101111011000110110101011
d9a5b9	5cdc99
inverse 0 and 1	
110110011010010110111001	010111001101110010011001
265a46	a32366

Fig. 5. Swapping of bits

generated bit values were weighted according to its hexadecimal values. These values are the equivalent encrypted text of the plaintext recognized by the user.

Figure 6 shows the equivalent encrypted text of the plaintext supplied by the user. The plaintext value bicolunivers contains the encrypted text bd8dab91a5b1265 a46a32366. Thus, the plaintext length was doubled after passing through the cycle operation using BCD 8421 code.

PlainText: bicolunivers | Encrypted Text: bd8dab91a5b1265a46a32366

Fig. 6. Encrypted text

3.2   Decryption

The 24-Byte length encrypted text was divided into 6 (six) characters per block. The character will be converted into its binary equivalent following the BCD 8421 con-version process.

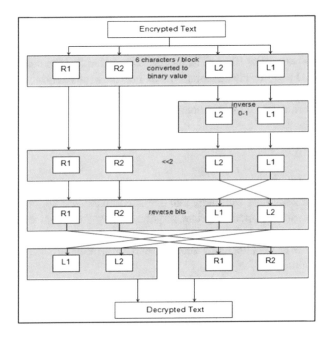

Fig. 7. Decryption process

Figure 7 shows the decryption process including the block cycle, swapping and Byte splitting operations. The decryption process follows the reverse order operation of the encryption process to obtain the equivalent ciphertext.

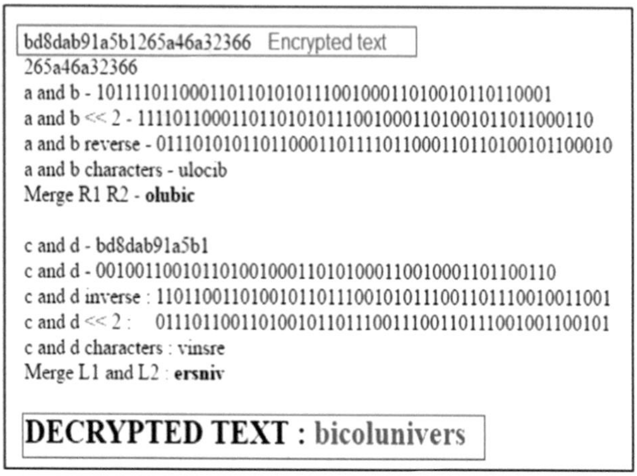

```
bd8dab91a5b1265a46a32366   Encrypted text
265a46a32366
a and b - 1011110110001101101010101110010001101001011011001
a and b << 2 - 1111011000110110101010111001000110100101101100110
a and b reverse - 0111010101101100011011110110001101101001011000010
a and b characters - ulocib
Merge R1 R2 - olubic

c and d - bd8dab91a5b1
c and d - 0010011001011010010001101010001100100011011001110
c and d inverse : 1101100110100101101110010101110011011110010011001
c and d << 2 :   0111011001101001011011100111001101110011001001100101
c and d characters : vinsre
Merge L1 and L2 : ersniv
```

DECRYPTED TEXT : bicolunivers

Fig. 8. Simulation of Encryption/Decryption process

Figure 8 shows the simulation of encryption/decryption processes how the decrypted value bd8dab91a5b1265a46a32366 was reverted back to its original value bicolunivers. The values generated by L and R divisions in each cycle operations especially the movement of BitShift to its desired location and the conversion of these values from BCD code to hexadecimal values provides satisfaction to the decryption process. Changing the location of the bits and characters with the inclusion of rotation and random insertion of key pattern helps the ciphertext become more complex.

4   Conclusion

This paper presents a new cipher technique that can reduce, save time processing encryption operation, offers simplicity and extended mixed encryption key generation. Cryptography is an important mechanism of securing important data. One of the requirements to accomplish is to secure the transmitted data over an unsecured communication channel. This includes security and protection to data via encryption/decryption method.

In this paper, a new technique applied with different bit-byte manipulation such as byte splitting, block size transposition, shift row and mix rows method. Every work has its own strength and weaknesses, scope and limitation. The combination of these techniques provides simple yet complex cipher for securing data. The new algorithm showed that there is sound interestingness behind the combination of the different techniques and that the cipher scheme is a potential to use.

Acknowledgment. This work (Grants No. C0515862) was supported by Business for Cooperative R&D between Industry, Academy, and Research Institute funded Korea Small and Medium Business Administration in 2017. Thank you to Bicol University, Technological Institute of the Philippines for the support given.

References

1. History of Encryption. SANS Institute of InfoSec Reading Room. Copyright SANS Institute. https://www.sans.org/reading-room/whitepapers/vpns/history-encryption-730
2. Rhee, M.Y.: Internet Security. Crytographic principles, algorithms, and protocols, p. 57
3. Data Encryption Standard. http://www.tutorialspoint.com
4. Playfair Cipher. http://practicalcryptography.com/ciphers/playfair-cipher/#references
5. Introduction to Blowfish. SplashID Safe Password Vault. http://www.splashdata.com/splashid/blowfish.htm
6. Schneier, B.: The Blowfish Encryption Algorithm. https://www.schneier.com/cryptography/blowfish/
7. BLOWFISHENC: Blowfish Encryption Algorithm. iitd.vlab.co.in
8. Internet Security, Chap. 4 [ppt] slide 45–46
9. Xijun LI, A Summary of Bit Manipulation. https://xijunlee.github.io/2017/04/01/efficiently/
10. Bansal, S., Shrivasta, R.: The IUP Journal of Computer sciences, vol. V, no. 4 (2011)
11. Bhagat, P., Satpute, K., Palekar, V.: Reverse encryption algorithm: a technique for encryption & decryption. Int. J. Latest Trends Eng. Technol. (IJLTET) 2(1), 91 (2013)
12. Lee, S., Park, T., Lee, D., Nam, T., Kim, S.: Chaotic order preserving encryption for efficient and secure queries on databases. IEICE Trans. Inf. Syst. **92**, 207–217 (2009)
13. Ceselli, A., Damiani, E., Vimercati, S.D.C.D., Jajodia, S., Paraboschi, S., Samarati, P.: Modeling and assessing inference exposure in encrypted databases. ACM Trans. Inf. Syst. Secur. **8**(1), 119–152 (2005)
14. Damiani, E., Di Vimercati, S.D.C., Finetti, M., Paraboschi, S., Samarati, P., Jajodia, S.: Implementation of a storage mechanism for untrusted DBMSs. In: IEE Security in Storage Workshop 2003, pp. 38–46 (2003)
15. Hacigümüş, H., Iyer, B., Mehrotra, S.: Ensuring the integrity of encrypted databases in the database-as-a-service model. In: DBSec 17th Annual Working Conference on Data and Application Security, Kluwer, pp. 61–74 (2003)
16. Tang, Q., Ji, D.: Verifiable attribute based encryption. Int. J. Netw. Secur. **10**(2), 114–120 (2010)
17. Kumar, Y., Joshi, R., Mandavi, T., Bharti, S., Rathour, R.: Enhancing the security of data using DES algorithm along with substitution technique. Int. J. Eng. Comput. Sci. **5**(10), 18395–18398 (2016). www.ijecs.in, ISSN 2319-7242
18. Jain, A., Dedia, R., Ppatil, A.: Enhancing the security of caesar cipher substitution method using a randomized approach for more secure communication. Int. J. Comput. Appl. **129**(13), 6–11
19. Vyas, B., Vajpayee, A.: Local data security thought encryption. IJSART **2**(8), 10–15 (2016)
20. Alanazi, H.O., Zaidan, B.B., Zaidan, A.A., Jalab, H.A., Shabbir, M., Al-Nabhani, Y.: New comparative study between DES, 3DES and AES within Nine factors. J. Comput. **2**(3), 152–157 (2010)
21. Ebrahim, M., Khan, S., Khalid, U.B.: Symmetric algorithm survey: a comparative analysis. Int. J. Comput. Appl. **61**(20), 12–19 (2013). https://arxiv.org/ftp/arxiv/papers/1405/1405.0398.pdf
22. Fathy, A., Tarrad, I.F., Hamed, H.F.A., Awad, A.I.: Advanced encryption standard algorithm: issues and implementation aspects. In: Hassanien, A.E., Salem, AB.M., Ramadan, R., Kim, T. (eds.) Advanced Machine Learning Technologies and Applications, AMLTA 2012. Communications in Computer and Information Science, vol 322. Springer, Heidelberg (2012)

23. Daemen, J., Rijmen, V.: The block cipher Rijndael. In: Smart Card Research and Applications. LNCS, vol. 1820, pp. 288–296. Springer
24. Pitchaiah, M., Daniel, P., Praveen: Implementation of advanced encryption standard algorithm. Int. J. Sci. Eng. Res. 3(3), 1–6 (2012). https://doi.org/10.15662/IJAREEIE.2016. 0506068
25. Katkade, P., Phade, G.M.: Application of AES algorithm for data security in serial communication. In: IEEE International Conference on Inventive Computation Technologies (ICICT), Coimbatore, India, August 2016
26. Binary Coded Decimal. ElectronicsTutorials. https://www.electronics-tutorials.ws/binary/ binary-coded-decimal.html

Reciprocating Link Hierarchical Clustering

Eric Goold$^{(\boxtimes)}$, Sean O'Neill, and Gongzhu Hu

Department of Computer Science, Central Michigan University,
Mount Pleasant, MI 48859, USA
{goold1eb,hu1g}@cmich.edu, oneil2sp@gmail.com

Abstract. A new clustering algorithm, called *reciprocating link hierarchical clustering*, is proposed which considers the neighborhood of the points in the data set in term of their reciprocating affinity, while accommodating the agglomerative hierarchical clustering paradigm. In comparison to six conventional clustering methods, the proposed method has been shown to achieve better results with cases of clusters of different sizes and varying densities. It successfully replicates the results of the mutual k-nearest neighbor method, and extends the capability to agglomerative hierarchical clustering.

Keywords: Clustering · Affinity · Reciprocating Link

1 Introduction

Clustering is a process of discovering groups (clusters) in a set of data records. The grouping is based on some metrics that measure the proximity (similarity or dissimilarity) among the data records. The goal is to generate clusters such that the average dissimilarity of data records in the same cluster is minimized while the average dissimilarity of data records in the different clusters is maximized. To achieve this goal, quite a number of clustering algorithms have been developed along with various proximity measures. These algorithms can be roughly categorized into two types: partitioning and hierarchical.

While clustering algorithms can perform very useful cluster analysis, sometimes they are unable to make the optimal clusters. For example, the single link hierarchical algorithm may result in clusters that can be tainted by the introduction of outliers that are close to another unrelated cluster. The k-means's clusters cannot be used if there is an extreme outlier introduced. Commonly used hierarchical clustering methods cannot handle regions of varying size or unusual shapes. In this paper, we propose a clustering algorithm that defines the borders of clusters by clustering data points based upon a shared *reciprocating affinity* with each other aiming at improving the quality of the clustering results. In this method, we define a similarity measure called *affinity* that measures the relative position with respect to its neighbors.

© Springer Nature Switzerland AG 2019
R. Lee (Ed.): ACIT 2018, SCI 788, pp. 149–165, 2019.
https://doi.org/10.1007/978-3-319-98370-7_12

Clustering methods that consider mutual affinity among data points has been studied for a long time. For example, the mutual k-nearest neighbor clustering, introduced in the early 1970's [14], is a classical method considering mutual affinity. Researchers have introduced various affinity metrics, such as clustering coefficient [26] and its variations, to measure the relationships between objects in social and communication networks for discovering clusters in the networks. We shall discuss some more previous work about clustering using mutual affinity information in Sect. 5.

The reciprocating affinity link clustering algorithm proposed in this paper is for hierarchical clustering that uses affinity ratio along with selected penalty functions as a merging condition. Several experiments were conducted using the proposed method. The results were evaluated with comparison to the results produced by several traditional algorithms. It shows that our algorithm generates better clusters for various data distributions than these traditional algorithms.

2   Background

We first briefly review the commonly used clustering algorithms for the purpose of comparison with the one we propose in this paper. The algorithms include the hierarchical clustering, k-means, DBSCAN, as well as mutual k-nearest neighbors that utilizes the neighborhood distribution of the data points.

2.1   Hierarchical Clustering

Hierarchical Clustering is a common way to perform cluster analysis. The basic idea of hierarchical clustering methods is to group data objects into a tree of clusters, viewed as a dendrogram. At the root level, there is only one cluster containing all data objects. At the leaf level, there are n clusters each of which contains only one data object. At an intermediate level, there are k clusters where k is the number of tree edges at that level. Hierarchical Clustering can be classified under two categories: agglomerative and divisive. The agglomerative approach starts from the leaves of the tree and moves up to the root, merging pairs of clusters into larger ones along the way. The divisive method starts from the root of the tree and moves down to the leaves, spiting larger clusters into smaller ones. Detailed discussion about hierarchical clustering can be found in some classical articles [5,16,20,24].

The basic hierarchical clustering algorithms are distinguished by the proximity they use to measure the similarity between clusters. In the following, we use $p(C_i, C_j)$ to denote the proximity between cluster C_i and C_j, and $d(x, y)$ for the "distance" (dissimilarity) of data points x and y. For a set of n data points, the distances can be represented as a $n \times n$ matrix D.

2.1.1 Single Link

A commonly used clustering method is single link clustering. In this method, the proximity of two clusters C_i and C_j is defined as the *minimum* distance of any two points $x \in C_i$ and $y \in C_j$.

$$p(C_i, C_j) = \min d(x, y), x \in C_i, y \in C_j \qquad (1)$$

Single link clustering is simple and performs well with well-separated data points. However, it is sensitive to outliers and not suitable for data sets that are stretched out.

2.1.2 Complete Link

Complete link clustering functions very similarly to single link clustering, but with a different proximity function that is defined as

$$p(C_i, C_j) = \max d(x, y), x \in C_i, y \in C_j \qquad (2)$$

That is, the proximity of two clusters C_i and C_j is defined as the *maximum* distance of any two points $x \in C_i$ and $y \in C_j$.

Complete link clustering, unlike single link clustering, can perform well with the presence of outliers in the data It does, however, have the likelihood of breaking up a large cluster into smaller clusters [13, 20].

2.1.3 Average Link

The average link clustering algorithm takes a middle way between single link and complete link. The proximity of two clusters C_i and C_j is the average proximity of all pairs of points in the different clusters, rather than the minimum or maximum.

$$p(C_i, C_j) = \frac{\sum_{x \in C_i, y \in C_j} p(x, y)}{|C_i| |C_j|} \qquad (3)$$

Average link clustering handles outliers better than single link because the presence of one point that is close to an outlier will not guarantee that point will merge. It suffers from the same shortcomings as complete link clustering.

2.1.4 Ward's Clustering

In Ward's clustering, the proximity of two clusters $p(C_i, C_j)$ is the increase of SSE (sum of squared errors) when two clusters are merged. It is similar to average method if $p(x, y) = d^2(x, y)$.

2.2 K-Means

K-means clustering [8, 11, 19] is a partitioning method to divide the data set into k clusters for a given k. k data points are selected as the initial centroids

$c_1, \cdots, c_k$ of the clusters, and other data points are assigned to the k clusters based on the proximity function. Point p is assigned to cluster i if

$$d(p, c_t) = \min_i d(p, c_i) \qquad (4)$$

New centroids are then recalculated and cluster memberships are reassigned until there is no change to the membership identification for any of the data points.

K-means, when compared to hierarchical clustering has a few pros and cons, is much more efficient as it does not build a hierarchy of clusters. However, k-means is sensitive to outliers and the selection of the initial centroids are critical. Different initial centroids may result in quite different clusters for the same data set, and some of which may be skewed. K-means excels when the clusters themselves are spherical.

2.3   DBSCAN

DBSCAN [7, 23] is a density based clustering method based on the concept of density-reachability. The density for point p in the data set is estimated as the number of points in a neighborhood of p with two parameters

(1) Eps: the radius of the neighborhood $N_{Eps}(p)$
(2) $MinPts$: the minimum number of points in the neighborhood

The two parameters define the Eps-neighborhood of p as

$$N_{Eps}(p) = \{q, d(p, q) < Eps\} \qquad (5)$$

A point p is *directly density-reachable* from point q if $p \in N_{Eps}(q)$ and $N_{Eps}(q) \geq MinPts$. A point p is *density-reachable* from point q if there is a chain of points $p_1, p_2, \cdots, p_k, p_1 = p, p_k = q$ such that p_{i+1} is directly density-reachable from p_i. A cluster identified by the DBSCAN algorithm is a maximum set of density-connected points from a given point. Any point that is not included in any cluster is considered an outlier.

DBSCAN is suitable for data sets where points of different clusters (Gestalt clusters) interweave with each other. It is less sensitive to noise, but does not perform well for overlapping clusters with different densities.

2.4   Mutual K-Nearest Neighbors

Different from the clustering algorithms mentioned above, Mutual K-Nearest Neighbors (MKNN) method considers two points to be assigned in the same cluster only when they are mutually close to each other. The idea of clustering based on mutual neighbors has been used for many years [4, 10, 12] that defined various similarity measures based on shared near neighbors. That is, this approach introduced an affinity idea to the MKNN method. The MKNN clustering algorithm is able to discover clusters of different densities and these clusters may overlap.

The MKNN clustering algorithm starts by creating candidate clusters using the MKNN property of the data points, and then repeatedly merges the candidate clusters until it has reached the predefined number of clusters or all pairs of the remaining clusters have zero connectivity.

3   Reciprocating Link

We propose a clustering method that has all the advantages of an agglomerative hierarchical clustering method with the capability of handling interesting shaped clusters like DBSCAN and mutual k-nearest neighbor clustering. The method is called reciprocating link clustering and accomplishes this by introducing a measure of similarity, called *affinity*, that is dependent upon each point's neighborhood within the data set. Due to the relative nature of the affinity measure,

---

Algorithm 1. Reciprocating Link

Input: P — point array
Input: $Dmat$ — distance matrix
Output: $Hierarchy$ — collection of clusterings

1 **begin**
2   **for** $p_1 : P$ **do**
3   determine $Rank(p_1, p_2), \forall p_2 \neq p_1$
4   **for** $p_2 \neq p_1 : P$ **do**
5   Calculate $ARmat(p_1, p_2)$ using Eq. (6)
6              $Amat(p, q) = F(Rank(p, q)) * ARmat(p, q)$
7   **end**
8   **end**
9   **for** $p_1 : P$ **do**
10   **for** $p_2 \neq p_1 : P$ **do**
11   Symmetrizing $Amat(p_1, p_2)$ using Eq. (11)
12   **end**
13   **end**
14   **for** $p : P$ **do**
15   put p into singleton cluster C_p
16   put C_p into $Clusters$
17   **end**
18   **while** $Clusters.size > 1$ **do**
19   **for** $C_1 : Clusters$ **do**
20   **for** $C_2 : Clusters$ **do**
21   find two clusters C_a and C_b with highest affinity
22   **end**
23   **end**
24   join cluster C_a and cluster C_b
25   put copy of $Clusters$ in $Hierarchy$
26   **end**
27   return $Hierarchy$
28 **end**

the resulting matrix is asymmetric, in contrast to the distance matrix typically used. This matrix is symmetrized, by assigning each value to the minimum of the opposing pair of values to which it belongs. Hence, the name, "reciprocating link" – the method clusters based upon the measure of affinity that is reciprocated between a pair of points. Algorithm 1 provides the basic procedure of the method.

In the algorithm, lines 2–8 iterate through all pairs of points to determine neighbor rankings and calculate the affinity ratio with a selected penalty function. The affinity matrix is made symmetric in lines 9–13, and initial singleton clusters are created in lines 14–17. The remaining part of the algorithm (lines 18–26) builds the clustering hierarchy that is returned on line 27. Details of each of these are explained in the sub-sections below.

3.1   Ranking Neighbors

The similarity measure, affinity, is relative to each point's position with regard to its neighbors. For each point, each neighbor is ranked according to its distance from the point. The closest neighbor is given a ranking of 1. In our method, the simple ordinal ranking is used, with ties being broken randomly. Other ranking strategies can be used instead. The ranking is later used by a penalty function which provides further differentiate affinities among neighbors.

3.2   Affinity Ratio

The affinity ratio is an intermediate measure of similarity, and is calculated by

$$ARmat(p_1, p_2) = \frac{min_{q \neq p_1}(Dmat(p_1, q))}{Dmat(p_1, p_2)} \tag{6}$$

The affinity ratio is left undefined for the diagonal entries in the matrix, and the off-diagonal entries have a value in the interval $(0, 1]$. In each row, the value of 1 corresponds to the closest neighbor to the point corresponding to the row.

3.3   Penalty Function

The penalty function is used to give further differentiation to affinity of a point towards it neighbors. A penalty function, F, must have the following properties:

- $F : N \rightarrow [0, 1]$
- $F(1) = 1$
- F is a nonincreasing function

A constant penalty function can be used if no further differentiation is needed. Figure 1 gives an illustration of the constant penalty function in comparison with the inverse penalty function, and exponential penalty functions:

$$Const(k) = 1$$

$$Inverse(k) = \frac{1}{k}$$

$$Exponential(k) = \frac{1}{2^{k-1}}$$

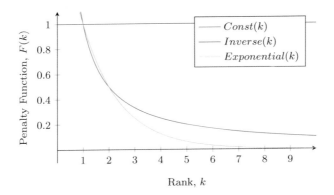

Fig. 1. Constant, inverse, and exponential penalty functions

The penalty function is applied to the affinity ratio to give the affinity measure according to

$$Amat(p, q) = F(Rank(p, q)) * ARmat(p, q).$$

If the penalty function has the value of 0, it will produce an affinity of 0. In our implementation of the method, clusters with zero affinity are not joined, as this clustering would have no basis and be arbitrary. Thus, a penalty function with a zero value will not produce a hierarchical clustering, rather the clustering will stop short.

The reciprocating link method produces the same results as the mutual k-nearest neighbor clustering if a suitable step function is employed as the penalty function. The step function is parameterized with a parameter, c, representing the cut-off ranking as given in Eq. (7).

$$Step_c(k) = \begin{cases} 1 & k \leq c \\ 0 & k > c \end{cases} \tag{7}$$

To improve upon the mutual k-nearest neighbor clustering behavior, it may be useful to provide the penalty function with an interval of more gradual decrease. A simple implementation of the concept is the ramp penalty function given in Eq. (8). This function takes two parameters: c, the position of the center of the ramp, and w, the width of the ramp.

$$Ramp_{c,w}(k) = \begin{cases} 1 & k \leq c - \frac{w}{2} \\ \frac{c-k}{w} + \frac{1}{2} & c - \frac{w}{2} < k < c + \frac{w}{2} \\ 0 & k \geq c + \frac{w}{2} \end{cases} \tag{8}$$

The ramp penalty function still has the drawback that it does give a value of zero beyond a certain rank. It would be better if the function approached zero as a limit but never reached it. What is desired is a function with two plateaus. The logistic function used by Verhulst [25] to describe population growth fits this criteria:

$$p = \frac{mp_o e^{mt}}{np_o e^{mt} + m - np_o} \tag{9}$$

Verhulst used three parameters: p_o, the initial population, m, the initial growth rate, and n, where the limiting population is m/n. For the purpose of a penalty function, we use two parameters: c, the ranking where it is approximately 0.50, and w, the width over which the penalty function drops from approximately 0.80 to 0.20 as given in Eq. (10). The graph of a logistic function is given in Fig. 2 in contrast to comparable step and ramp penalty functions. The use of the logistic penalty function adds fluidity to the mutual k-nearest neighbor method along with extending it with the capability of agglomerative hierarchical clustering.

$$Logistic_{c,w}(k) = \frac{1 + e^{\frac{2.77(1-c)}{w}}}{1 + e^{\frac{2.77(k-c)}{w}}} \tag{10}$$

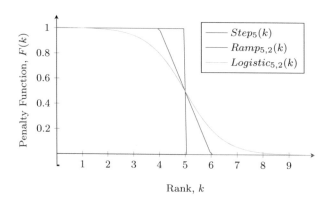

Fig. 2. Step, ramp, and logistic penalty functions

3.4   Symmetrizing Affinity Matrix

Finally, before actual clustering, the affinity matrix must be symmetrized. This is done by the assignment in Eq. 11.

$$Amat(p_1, p_2) = min(Amat(p_1, p_2), Amat(p_2, p_1)) \tag{11}$$

This achieves the reciprocity of affinity in the reciprocating link method.

3.5   Hierarchical Clustering

The actual clustering follows exactly like the single link hierarchical clustering. However, instead of finding the minimum of distances, which is a measure of separation, this method finds the maximum of affinities, which is a measure of similarity. If a tie is met, it is broken randomly. In our implementation, if zero affinity is found between clusters, it does not join them, and the method will not reach the completion of a single cluster containing all points. This is done by choice, so that the results of the mutual k-nearest neighbor can be replicated. The method can easily be altered to include arbitrary joining of clusters until the hierarchical clustering is complete.

4   Experimental Results

To determine the performance of the reciprocating link clustering, we implemented it in Java 1.8 using the JFreeChart [15] package for visualization purposes. The data set we used for testing is shown in Fig. 3. This data set was contrived to test the capability of clustering a small dense configuration within a large sparse cloud.

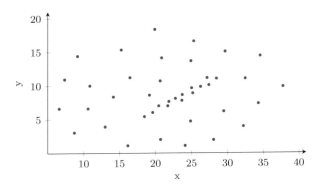

Fig. 3. Unclustered data set

4.1   Conventional Clustering Results

The six conventional clustering methods were used on the data set, and the results are in Fig. 4. In Single Link Clustering, due to border points getting clustered into the central cluster, every point but one was clustered together as shown in Fig. 4.

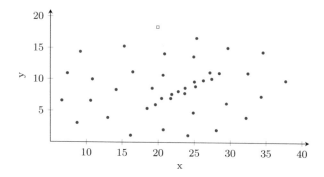

Fig. 4. Clustering result of single link

In Complete Link Clustering, the algorithm could not differentiate between the central and outer clusters and instead clustered the points into a left and right cluster, as shown in Fig. 5.

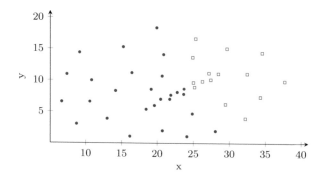

Fig. 5. Clustering result of complete link

The clustering result of Average Link Clustering is shown in Fig. 6. It could not differentiate between the central and outer cluster as well, but the entire central cluster was contained into one cluster along with many points in the outer cluster.

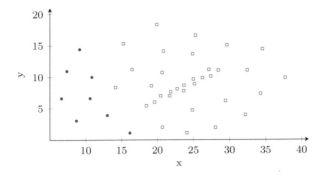

Fig. 6. Clustering result of average link

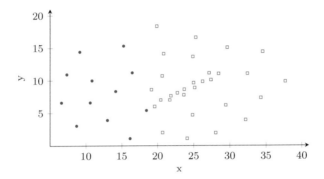

Fig. 7. Clustering result of K-Means

K-Means clustering was not able to sort the points into their two proper clusters and instead split the points down the middle into two clusters, as shown in Fig. 7.

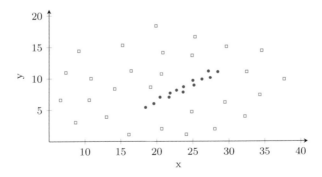

Fig. 8. Clustering result of DBSCAN

Using DBSCAN with $Eps = 2.0$ and $MinPts = 3$ clusters the central group successfully, separating it from the greater sparse cloud, as shown in Fig. 8.

Mutual K-Nearest Neighbors was unable to even sort down to two clusters and instead was left with six clusters. While the entire central cluster was accurately identified as a single cluster, the algorithm struggled to combine all the surrounding clusters. The clustering result of Mutual K-Means is shown in Fig. 9.

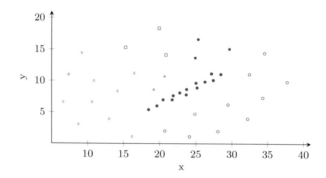

Fig. 9. Clustering result of mutual K-Means

4.2   Reciprocating Link Results

The results of the reciprocating link clustering is given in Fig. 10. Notice that when using a constant penalty function, the method did not achieve the desired results. However, it is better than the other hierarchical methods: single link, complete link, and average link. Using the inverse penalty function caused more pronounced distinction among the neighbors and successfully clustered the data set. The exponential and logistic penalty functions were also successful giving the same results.

4.3   Multi-density Results

To further test the capabilities of the reciprocal link clustering method, the data set in Fig. 11 was used. This data set was generated randomly under the condition that points within each of the three distinct regions were allowed a different minimum separation from all other points. The motivation was to test the capability of the method to differentiate between densities of three different levels. This is a task beyond the capabilities of DBSCAN, which performed so well on the previous data set.

The mutual k-nearest neighbor clustering algorithm was able to keep the three density regions distinct with $k = 3$. However, it was unable to cluster the regions as individual clusters as shown in Fig. 12. The reciprocating link method with the $Step_3$ penalty function achieved the exact same clustering,

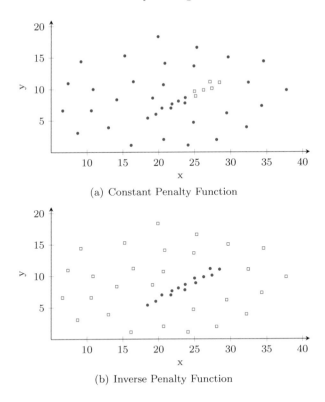

(a) Constant Penalty Function

(b) Inverse Penalty Function

Fig. 10. Reciprocating link clustering

which demonstrates its capability of reproducing the results of the mutual k-nearest neighbor algorithm.

The reciprocating link method did better using the $logistic_{5,4}$ penalty function, as shown in Fig. 13. With 8 clusters, the three density regions are separately

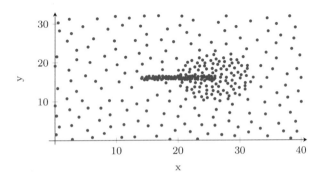

Fig. 11. Unclustered data set

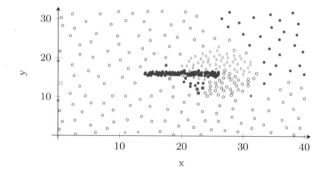

Fig. 12. Mutual K-Nearest neighbors

clustered as individual clusters with the exception of five singleton clusters. With 3 clusters, the three density regions are, unfortunately, combined into a single cluster while two stubborn singleton clusters remain.

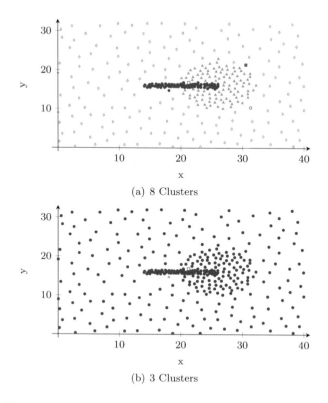

(a) 8 Clusters

(b) 3 Clusters

Fig. 13. Reciprocating Link with $Logistic_{5,4}$ penalty function

The persistence of singleton clusters, with exceptionally low affinities, parallels the problems with outliers in the single link clustering. The points with very low affinities behave like the outliers with very high distances. However, the difference is that in the reciprocating link method, these occur on the boundaries of natural clusters, while in the single link, they are typically located at the fringes of the data set.

5   Related Work

The issue of reciprocating affinity among objects is commonly discussed in social study and network analysis, particularly social network (or communication network) analysis, where finding the mutual relationships is often the goal. Studies have shown that communication networks tend to form links between neighbors (triangles) deviating from random networks. Such tendency reflects the affinity among nodes to form clusters. One of the commonly used measures is clustering coefficient [26]. Saramäki et al. [22] presented a study of various weights of links in social networks to generalize the clustering coefficient measure of clustering of network nodes. The cluster coefficient of node i is defined as

$$C_i = \frac{2t_i}{k_i(k_i - 1)}$$

where k_i is the degree of node i, and t_i is the number of edges among i's neighbors. The cluster coefficient of the network is the average of C_i of all node i in the network. Yusheng Li et al. provided a theoretical study on cluster coefficient [18] on large networks. Measures like cluster coefficients have been extensively researched to analyze the behaviors of social networks.

Rong et al. showed how the cluster coefficient influences collaborations on highly clustered networks [21]. They indicated that "triangle loops provide stronger support for mutual cooperation" through the feedback reciprocity mechanism. Engel studied the effects of emailing in email communication networks on clustering and reciprocity [6]. Their findings via experiments and analysis showed that emails with multiple-recipients have high influence on forming dense clusters than emails with single recipient. A new method for studying clustering in networks was proposed by Berehaut et al. [2], in which the cluster coefficient was generalized when using random walk pairs for clustering.

A minimum spanning tree (MST) based clustering method, also called affinity, was proposed by Bateni et al. [1]. It is a hierarchical clustering on a very large scale based on a massively parallel MST algorithm and distributed hash tables. Experiments show that the algorithms is scalable to handle network with billions of nodes and trillions of edges.

Another clustering method using affinity information is affinity propagation (AP) introduced by Frey and Dueck in 2007 [9]. Since then, many algorithms have been developed based on the idea of affinity propagation, such as [17], including AP implementations provided in some analysis tools such as APCluster in R [3]. AP algorithms are basically partitioning rather than hierarchical.

The basic idea of AP is to derive a set of "exemplar" data points as the centroids for clustering by passing real-valued messages along the edges of the network of points. The messages are defined by formulas that minimize an energy function, and the magnitude of the message is a measure of the current affinity among data points. Our approach, however, is quite different from the AP algorithms that it is for hierarchical clustering and hence no need to select exemplar points for partition, and affinity is a similarity measure rather than a message being propagated.

The reciprocating link method proposed in this paper is to some extent similar to mutual k-nearest neighbor, but extended to hierarchical clustering with various penalty functions incorporated in the algorithm.

6   Conclusion

In comparison to the six conventional clustering methods, the reciprocating link method has been shown to achieve better results with cases of clusters of different sizes and different densities. The reciprocating link successfully replicates the results of the mutual k-nearest neighbor method using a step penalty function, and extends the capability to agglomerative hierarchical clustering when using a logistic penalty function, while achieving better results. An unexpected flaw in the reciprocating link method is the development of persistent singleton clusters at the border of the natural clusters. Further development, perhaps through the introduction of a complimentary mechanism targeted at the persistent singletons, may solve this problem in the future. Further tests on real data sets are also needed for a complete comparison of its performance with conventional methods.

References

1. Bateni, M., Behnezhad, S., Derakhshan, M., Hajiaghayi, M., Kiveris, R., Lattanzi, S., Mirrokni, V.S.: Affinity clustering: hierarchical clustering at scale. In: Advances in Neural Information Processing Systems 30: Annual Conference on Neural Information Processing Systems, pp. 6867–6877, Long Beach, CA, USA (2017)
2. Berenhaut, K.S., Kotsonis, R.C., Jiang, H.: A new look at clustering coefficients with generalization to weighted and multi-faction networks. Soc. Netw. **52**, 201–212 (2018)
3. Bodenhofer, U., Kothmeier, A., Hochreiter, S.: ApCluster: an R package for affinity propagation clustering. Bioinformatics **27**(17), 2463–2464 (2011)
4. Brito, M., Chávez, E., Quiroz, A., Yukich, J.: Connectivity of the mutual k-nearest-neighbor graph in clustering and outlier detection. Stat. Probab. Lett. **35**(1), 33–42 (1997)
5. Day, W.H.E., Edelsbrunner, H.: Efficient algorithms for agglomerative hierarchical clustering methods. J. Classif. **1**, 7–24 (1984)
6. Engel, O.: Clusters, recipients and reciprocity: extracting more value from email communication networks. Procedia Soc. Behav. Sci. **10**, 172–182 (2011)

7. Ester, M., Kriegel, H.P., Sander, J., Xu, X.: A density-based algorithm for discovering clusters in large spatial databases with noise. In: Proceedings of the 2nd International Conference on Knowledge Discovery and Data Mining, pp. 226–231. AAAI Press, Menlo Park (1996)
8. Forgey, E.: Cluster analysis of multivariate data: efficiency vs. interpretability of classification. Biometrics **21**, 768 (1965)
9. Frey, B.J., Dueck, D.: Clustering by passing messages between data points. Science **315**, 972–976 (2007)
10. Gowda, K.C., Krishna, G.: Agglomerative clustering using the concept of mutual nearest neighbourhood. Pattern Recogn. **10**(2), 105–112 (1978)
11. Hartigan, J.A., Wong, M.A.: Algorithm as 136: a k-means clustering algorithm. J. R. Stat. Soc. Ser. C **28**(1), 100–108 (1979)
12. Hu, Z., Bhatnagar, R.: Clustering algorithm based on mutual k-nearest neighbor relationships. Stat. Anal. Data Mining **5**(2), 100–113 (2012)
13. Hubert, L.: Approximate evaluation techniques for the single-link and complete-link hierarchical clustering procedures. J. Am. Stat. Assoc. **69**(347), 698–704 (1972)
14. Jarvis, R., Patrick, E.: Clustering using a similarity measure based on shared near neighbors. IEEE Trans. Comput. C-22(11), 1025–1024 (1973)
15. Jfreechart. http://www.jfree.org/jfreechart. Accessed 2017
16. Lance, G.N., Williams, W.T.: A general theory of classificatory sorting strategies: 1. hierarchical systems. Comput. J. **9**(4), 373–380 (1967)
17. Li, P., Ji, H., Wang, B., Huang, Z., Li, H.: Adjustable preference affinity propagation clustering. Pattern Recogn. Lett. **85**, 72–78 (2017)
18. Li, Y., Shang, Y., Yang, Y.: Clustering coefficients of large networks. Inf. Sci. **382–383**, 350–358 (2017)
19. MacQueen, J.: Some methods for classification and analysis of multivariate observations. In: Proceedings of the Fifth Berkeley Symposium Mathematics, Statistics and Probability, vol. 1, pp. 281–296. University of California Press, Berkeley (1967)
20. Murtagh, F.: A survey of recent advances in hierarchical clustering algorithms. Comput. J. **26**(4), 354–359 (1983)
21. Rong, Z., Yang, H.X., Wang, W.X.: Feedback reciprocity mechanism promotes the cooperation of highly clustered scale-free networks. Phys. Rev. E **82**, 047,101 (2010)
22. Saramäki, J., Kivelä, M., Onnela, J.P., Kaski, K., Kertész, J.: Generalizations of the clustering coefficient to weighted complex networks. Phys. Rev. E **75**, 027,105 (2007)
23. Schubert, E., Sander, J., Ester, M., Kriegel, H., Xu, X.: DBSCAN revisited, revisited: why and how you should (still) use DBSCAN. ACM Trans. Database Syst. **42**(3), 19:1–19:21 (2017)
24. Sneath, P.H.A., Sokal, R.R.: Numerical Taxonomy. The Principles and Practice of Numerical Classification. W. H. Freeman, San Francisco (1973)
25. Verhulst, P.F.: Notice sur la loi que la population poursuit dans son accroissement. Correspondance mathématique et physique **10**, 113–121 (1838)
26. Watts, D.J., Strogatz, S.H.: Collective dynamics of 'small-world' networks. Nature **393**, 440–442 (1998)

Development of Infant Care System Application

Mechelle Grace Zaragoza and Haeng-Kon Kim[✉]

Daegu Catholic University, Gyeongsan-si, South Korea
mechellezaragoza@gmail.com, hangkon@cu.ac.kr

Abstract. Newborn infants spend up to 70% their time asleep. Neonates sleep at least 16–18 h per day, and their sleep patterns are markedly different from the sleep patterns of older infants and adults. Ideal sleep prepares the child to learn when she is awake, and after the learning has taken place during the awake state, the critical processes of memory association occur during sleep. Now Clearly, sleep deprivation, whether it's for an infant or parents who are involved needs to be addressed. The good news is that there are strategies you can use to get the rest you need. First, topics related to SIDS vulnerability, to better understand the role of learning during sleep in promoting infant survival are discussed. Second, the importance of making sure that infants are given the comfort they need in order to continuously sleep without disturbing the guardians. To address both issues of deprived sleep for infants and parents and Lasty, a mobile application for the system.

Keywords: Healthcare · Infant care system · Mobile application
SIDS

1 Introduction

The term "child care" refers to the social welfare service that provides support to nurseries and foster homes, care, and feeding of newborns from 0 to 5 years of age in a healthy and safe way, while offering them an education adapted to their needs which characteristics are of mental and physical development [1]. We know that infants and young children live the range of socio-emotional functions ranging from development that seems to be on the right path (for example, the ability to establish good relationships with others, play, communicate, learn and experience various emotions [2]. In order to fully understand the importance of child care, health monitoring became quite popular these days. Baby Monitor is an application that helps you control your baby even when you are not there. If your baby is sleeping and you are in another room, the baby monitor will detect if your baby is crying and will alert you by a call or a text message, will send alerts or will display notifications using modern technology devices [3].

1.1 Health Monitoring

Health monitoring is an essential application of portable detection systems, whether for adults or babies. Recently, client data available to hotel researchers and industry

© Springer Nature Switzerland AG 2019
R. Lee (Ed.): ACIT 2018, SCI 788, pp. 166–174, 2019.
https://doi.org/10.1007/978-3-319-98370-7_13

professionals or even medical related have increased significantly due to the emergence of the Internet of Things (IoT) [4] with advances in sensor technology, wireless communication technologies and power supply, portable sensor systems have allowed the creation of a new generation of continuous monitoring of children's health. However, conventional medical instruments and sensors cannot be used for portable physiological monitoring applications because they are difficult to use for long periods of time and also cause discomfort to the user. Due to technological limitations in sensors, wireless networks and power supply, conventional portable sensor systems are generally not suitable for robust, long-term and comfortable control of infants in real-life conditions. In addition, with increasing attention to the needs and clinical needs of babies, the application of integrated detection technology and consumer electronics improves the reliability and comfort of quality-assured surveillance systems. Portable electronics and smart textiles effectively prevent babies from being disrupted by conventional detection techniques that may include skin irritation, cable discomfort, sleep disruption and lack of communication with parents [5]. Caring for younger patients requires the best and most innovative technology. But, as we see, intensive neonatal therapy also requires something else: the focus on the NICU as a complete environment where babies can receive the therapies they need in a nurturing environment, but designed for flexibility and efficiency [6]. Even when they are asleep, babies are surrounded by environmental contingencies, which often cover several sensory modalities. For example, a decrease in heat reliably follows the removal of a diaper blanket. Could newborns know the environmental contingencies they experience during sleep, in addition to those experienced during their ephemeral periods of awakening? This question is of great interest to SIDS researchers because of its implications for child survival. Some of the situations encountered during sleep, such as respiratory occlusion and thermal problems, require a response from the baby, and babies who are already at risk for SIDS and have difficulty learning adaptively may be particularly vulnerable. Before exploring whether babies can learn an association between two sensory stimuli during sleep, it is necessary to establish (1) that newborns are capable of learning and (2) that they are capable of trying new information during sleep [7].

Not all parents can afford an excellent baby monitor, especially those sold at a price up to $300 or more. Never mind. If you have an Android or iOS device laying around catching dust, it's time to make full use of it and turn it into an iOS or Android baby monitor at a low cost, or even free. Baby monitor apps can be categorized into two main types: one is iPhone and Android phone apps that are compatible with baby monitors. The other is baby monitoring apps for iPhone, iPad, Android phone, which turn your smartphone into mobile baby monitors. I'll list several common baby monitoring apps for both iOS and Android devices. You'll never know how well they can work until you use by yourself. So, download and try these baby monitor apps while you are reading this article. Let's just get started [8].

2   Background of the Study

2.1   SIDS

SIDS is officially defined as the sudden death of a baby less than one year of age who remains unexplained after a thorough investigation, which includes the completion of a complete autopsy, examination of the scene of death and examination. The incidence of cases classified as SIDS by the National Center for Health Statistics has decreased since 1980. From 1980 to 2010, the rate decreased by 66%. Most of this decline occurred between 1990 and 2000, the decade after the US Return to sleep campaign, with a 57% decrease [9] (Fig. 1).

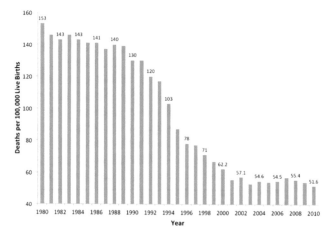

Fig. 1. SIDS Death per 100,000 Live Births 1980–2010

2.2   Role of Mobile Application

The first months of having a baby are rough, regardless of whether you are a repeat parent, like me, or a rookie just entering the exhilarating and exhausting fray. In those newborn months and even through the first year, most weary-eyed parents go into survival mode, yet miraculously adapt to all of the challenges that come with the transition. While there isn't any technology to get your baby to start sleeping through the night at eight weeks (yet!), there are plenty of great apps that can help you in one way or another during that foggy and fantastic first year.

2.2.1   Ready, Set, Baby!

Leave that tattered baby book on your nightstand and instead turn to this all-encompassing app or e-book, which is a great resource for every parent to have during the first year. Exactly how many diapers do babies go through when they're newborns? It might seem mundane to be tracking diapers, but what you're really tracking is output (and intake). Weight gain is crucial, and for nursing moms, especially, dirty and wet diapers are the only indication that babies are eating enough. Eat Sleep Tracker not only keeps tabs on diaper changes, but feeding times and sleep patterns, too.

2.2.2   Eat Sleep: Simple Baby Tracking

Exactly how many diapers do babies go through when they're newborns? It might seem mundane to be tracking diapers, but what you're really tracking is output (and intake). Weight gain is crucial, and for nursing moms, especially, dirty and wet diapers are the only indication that babies are eating enough. Eat Sleep Tracker not only keeps tabs on diaper changes, but feeding times and sleep patterns, too.

2.2.3   Best Baby Monitor

If you have an extra iOS device lying around, you can create an insta-baby monitor using this app. With one device in the nursery and one as the parent unit, Best Baby Monitor provides real-time video and audio of your baby using the same Wi-Fi network. You can even record those cute gurgles and coos and talk to your baby through the phone to shhh her back to a sound slumber.

2.2.4   Baby Soother

Speaking of sleep, any magical tips to actually get a baby to fall asleep are always welcomed by parents. People assume that the house has to be museum quiet when baby is sleeping, but in reality, the womb was quite noisy. Sound Sleeper is a great app to fill the silence with options such as ocean waves, white noise, hair dryer, vacuum and more.

2.2.5   BabyBook

Babies go through so many firsts, each more exciting than the last. BabyBook is a terrific app that lets you document all these milestones and preserve memories from your baby's everyday life. With easy-to-use templates, you can seamlessly fill up a book of moments from your mobile device that you can share with family and friends, and even print into a hardcover book.

2.2.6   iVaccine

The first year is filled with immunizations that are important for your baby's wellbeing. With several shots given at various pediatric appointments, it can be difficult to keep track of when your baby gets what. iVaccine keeps a handy log for your child (or multiple children), provides helpful information on each vaccine administered and even sends reminders of your child's upcoming vaccines.

2.2.7   Joya Video

With a baby in the house, nearly every moment seems worth recording. But that two minute-long video of your baby adorably trying to roll over is too long to send – or is it? With Joya, there is no video clip too large to send to family, friends, relatives and anyone else who would be interested. The app sends the video via email or text, and recipients can view the clip with a private link. Now, everyone can marvel at how your baby rolls.

2.2.8 Umano

With a new baby in the house, it can be difficult to find time to read for pleasure, let alone stay on top of current headlines. With Umano, the app that actually reads the news to you, you can still be feeding, burping, changing or bathing the baby while keeping up with current events. And with its use of professional voice actors, you'll feel like it's your own personal story time [10].

3 Mobile Application

The Proposed GUI for the application named NANNYDUINO as a remote controlled application for sample e-crib as discussed in Figs. 8, 9, 10, 11, 12, 13, 14 and 15 with details.

3.1 Proposed Mobile Application and Program

See Figs. 2, 3, 4, 5 and 6.

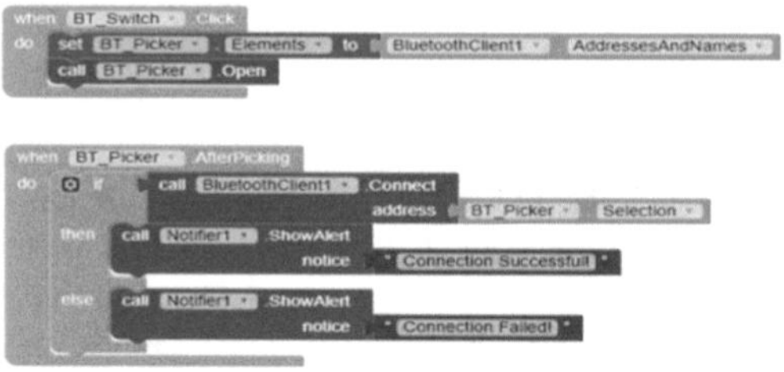

Fig. 2. Establish connection with bluetooth module

Fig. 3. Bluetooth message to send when buttons/switches are clicked

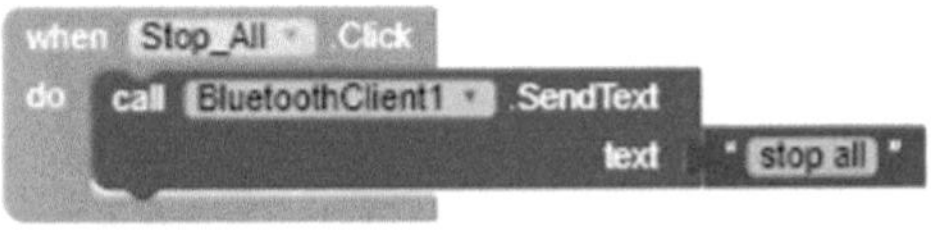

Fig. 4. Bluetooth message to send when emergency stop button is pressed

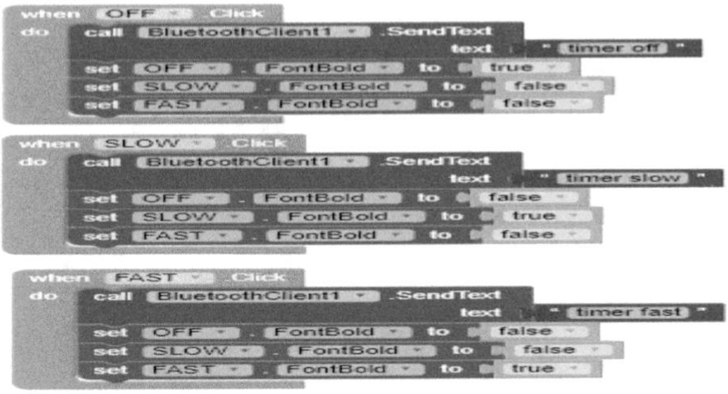

Fig. 5. Bluetooth message to send to choose among different timer speeds

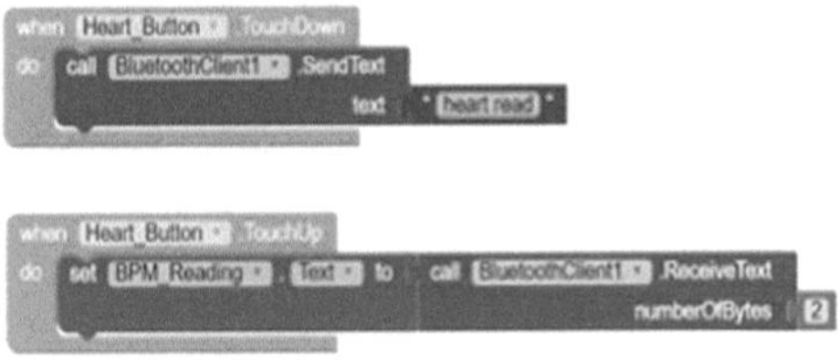

Fig. 6. Fetching heart reader

3.2 Application Design

See Fig. 7.

Fig. 7. Connecting bluetooth device

Fig. 8. Emergency stop/stop all devices

Fig. 9. Switching music off/on/play/next

Fig. 10. Turning fan on/off

Fig. 11. Swaying crib on/off

Fig. 12. Moving canopy up/down

Fig. 13. Selecting auto-off timer speed set to fast by default

Fig. 14. Fetching heart rate reading

Fig. 15. Heart beat reader

The sample proposed application design for are Ideal Icrib has the following button features:

Cheap E-Cradle
Would you like to have a swaying cradle?
Soothing music
Mobile App controller (remote control)
Wet sensor (For diaper and soil Detection)
Alarm sensor (For Crying babies)
Heart rate sensor
Mosquito canopy (automatic)
Fan (cooling fan attached)
Moving/rotating toy

4 Testing and Evaluation

Table 1, shows the steps in connecting the app via bluetooth to the system using an android phone.

Table 1. SKK mobile rush pixie android phone connection

Step 1. Pair the phone with the Bluetooth module.	Step 2. Enter the password. The default is "1234".	Step 3. Open the app.

5 Evaluation and Conclusion

Mobile applications, like any other software, have become more complex and should be constantly updated to: correct errors, add or remove functions, improve the user interface and more [11]. Today, babies can feel more comfortable with health control and doctors or guardians are immediately informed by mobile phone when health status is threatened. Once these developments are recognized, not only can babies receive ongoing medical care, but tutors and doctors can effortlessly understand a baby's needs. It will have favorable effects on the guardians and will be widely adopted in everyday life, various clinical situations and other possible applications of child supervision. There is a provision of command of the music to be played; motion of the fan and of the crib sway; a motion for the canopy and heart rate sensor. A baby's movement is not stable all the time considering that the hardware device has a wearable sensor. Apparently, this fact needs to be considered. Power management is an important factor in a wireless sensor network. It needs electric power to run. The system needs power to run the sensors and other hardware circuitry. The system may shut down if there is no presence of electricity.

Acknowledgements. This Research was supported by the MSIP (Ministry of Science, ICT and Future Planning), Korea, under the ITRC (Information Technology Research Center) support program (IITP-2018-2013-0-0087) supervised by the IITP (Institute for Information & Communications Technology Promotion).

References

1. Infant care. http://www.ilo.org/dyn/travail/docs/1764/Infant%20Care.pdf
2. Infant Mental Health and Early, Care and Education Providers. http://csefel.vanderbilt.edu/documents/rs_infant_mental_health.pdf
3. https://play.google.com/store/apps/details?id=dk.mvainformatics.android.babymonitor
4. Lee, S., Hwang, E., Jo, J.-Y., Kim, Y.: Big data analysis with hadoop on personalized incentive model with statistical hotel customer data. Int. J. Softw. Innovation (IJSI) **4**(3), 1–21 (2016)
5. Zhu, Z., Liu, T., Li, G., Li, T., Inoue, Y.: Wearable sensor systems for infants. Sensors **15**(2), 3721–3749 (2015)
6. Dräger Medical AG & Co. KG changes into Dräger Medical GmbH. Infant Intensive Care System, Dräger. Technology for Life, September 2010
7. Tarullo, A.R., Balsam, P.D., Fifer, W.P.: Sleep and infant learning. Infant child Dev. **20**(1), 35–46 (2011)
8. Li, A.: Top Baby Monitor Apps for iPhone/iPad/Android Devices, redlink.com, 26 September 2017
9. American SIDS Institute. https://sids.org/what-is-sidssuid/incidence/
10. Tahnk, J.L.: 1 0 Apps to Help You Through Baby's First Year, Mashable Asia, 5 February 2014
11. Saifan, A.A., Alsghaier, H., Alkhateeb, K.: Evaluating the understandability of android applications. Int. J. Softw. Innovation (IJSI) **6**(1), 44–57 (2018)

Author Index

A
Ahn, Kiljae, 48
Alqaydi, Lamya, 78

B
Balilo Jr., Benedicto B., 137
Bo, Chen, 104
Byun, Yungcheol, 137

C
Chun, Yang-Ha, 1

D
Damiani, Ernesto, 78
Donzia, Symphorien Karl Yoki, 37

G
Gerardo, Bobby D., 137
Gim, Gwang-yong, 104
Gim, Sang-Hoon, 48
Goold, Eric, 149

H
Hong, Ki-Seob, 124
Hu, Gongzhu, 149
Hwang, Ha Jin, 27

J
Jeon, Hyo Won, 89
Jo, Donghyuk, 62
Jung, Kyung-Jin, 104

K
Kim, Dong-Hyun, 124
Kim, Haeng-Kon, 14, 27, 37, 166
Kim, Hyo-Bin, 124
Kim, Jong-Bae, 1
Ko, Dae-Sik, 48

M
Marland, Veiga Yamodo, 14

O
O'Neill, Sean, 149

P
Park, Jung-Boem, 104
Phan, Nhu Quynh, 104

S
Seo, Jung-Taek, 124
Shin, Jeong Hoon, 89

Y
Yeo, Hyun, 37
Yeun, Chan Yeob, 78
Yoon, Soo-Yeon, 1

Z
Zaragoza, Mechelle Grace, 27, 166

© Springer Nature Switzerland AG 2019
R. Lee (Ed.): ACIT 2018, SCI 788, p. 175, 2019.
https://doi.org/10.1007/978-3-319-98370-7

Printed in the United States
By Bookmasters